Battle for the Solomons

BY
IRA
WOLFERT

 HOUGHTON MIFFLIN COMPANY · BOSTON
The Riverside Press Cambridge

PUBLISHED, JANUARY, 1943
SECOND IMPRESSION, JANUARY, 1943
THIRD IMPRESSION, JANUARY, 1943

The Riverside Press
CAMBRIDGE · MASSACHUSETTS
PRINTED IN THE U.S.A.

★

To the 26th Heavy Bombardment
Squadron — *their   present*

and future

★

If an Angel you would be
Please return this book to me.
Mildred Murtagh
Robertson

CONTENTS

CHAPTER 1

Nice Shooting Weather

★

CHAPTER 1

★

Nice Shooting Weather

SOMEWHERE ON THE PACIFIC

At about eleven o'clock in the morning, ship's time, one of the escort vessels smelled something and went to find it. It had been mousing along up ahead of the convoy, whiskers twitching, as they say to indicate that its mechanical feelers were spreading restlessly and combing out the water for dynamite. Now it gave up fooling and went. The bow threw up white water in a snarl, a whole snootful of it, and made the ship look as if it were lunging with teeth bared.

All the steering wheels in the convoy went hard to starboard. A big Danish water-tender was sitting on an empty egg crate in the engine room, whetting a knife sleepily. He looked like a coil of hawser piled up there. The sharp turn dumped him off the box and he fell, still sitting, on the iron deck. He fell lightly, his muscles being powerful enough almost to hold him up in midair, but his face was all round and like a child's with fright as he looked at

1

me trying to keep from winding up against anything hot. When he picked himself up, he muttered something about there being 'screwy nuts loose' in the wheelhouse nowadays with all the college boys in the Merchant Marine, and went back to whetting his knife.

The black gang, not unexpectedly, is worried about being where it is when a torpedo hits. This ship has Scotch marine boilers on it and all the fellows know it. A few might tend to forget it occasionally, but the first assistant, a squinty-eyed little man with a paunch on him that he calls his poop deck and ornaments with tattooing, keeps reminding them. Marine boilers, unlike fused boilers, do not simmer or cook when poked up by high explosives but just blow with a great big wham and there you are, without time even to say good-bye.

This first assistant tangled with the Japs early in the war before the freighters had anything to throw except signal flares. The Japs were conserving torpedoes that day. They put twenty-four three-inch shells into his ship, once blowing a lifeboat right out from under him as it was being swung out on the davits, and blowing him back on deck. He spent eleven days in an open boat with thirty-four men, two of whom died and two of whom went crazy and had to be sat on to be kept quiet. He's still got the

2

look in his eyes of a fellow watching muzzle-flashes in the distance and waiting for the roar and smack to tell him whether he's a dead pigeon or can still fly.

The oilers, wipers, firemen, and water-tenders all looked at me furtively as I walked along the narrow, open-sided companionway toward the ladder to topside. It was very quiet. Nobody said anything, but it was plain to see they hankered to go with me and see what was doing up above. I tried to walk calmly and deliberately. My heavy shoes rang against the iron way and the last thing I heard as I climbed the warm, greasy ladder was the first assistant helping a gang set in a flywheel on a refrigerator compressor and saying, 'Easy now; easy does it.' His voice sounded harsh and irritable.

It couldn't have been very long between the time the ship lurched heavily to starboard and the time I reached the deck. The ships of the convoy were still wheeling in station. They were, for the most part, old iron tubs with asthmatic engines that made them bubble and wheeze in every seam. They were all straining now. The glaring sun was as unkind to them as it is to all old girls. But there was something very gallant about the way they wheeled and kept stations and fluffed themselves up martially at the smokestacks, like old soldiers on parade,

3

trying to firm up and hide what they'd become. One of the ships was far out of station, as it had been nearly all the way from port. It was a steam-schooner, one of those thirty-day wonders from the last war. Its heart was in the right place, as it proved by joining up in this war, but its heart had leaky valves and it just couldn't keep from strag-gling out of every convoy. It had had its whole stern blown off once by a Jap torpedo and had made port, wallowing all the way and finally touching bottom in the harbor. Now its black gang was pour-ing it on. Black smoke was blasting out of its funnel and went streaking across the sky toward us, getting thinner and thinner until finally, when it reached us, it was as thin as a wail, like a distant, thinned-out cry of 'Wait for me!' Nobody was waiting.

'They must be having the conniptions there,' I said to the ordinary seaman standing next to me.

'I don't think so,' the man said. He's a fellow out of Lansing, Michigan, his studies having been detoured from college by the war. 'It seems to work out in the Merchant Marine and in the Navy that if you're the kind that gets conniptions you don't ship out.'

The escort vessel was still boiling away at some-thing. It was shimmering sullenly on the horizon of the brasslike sea, and you could see by its wake how fast it was going. Then it stopped, all hunkered down, whiskers twitching, feelers feeling, and stopped a long time and suddenly was off again. In a moment it was hull down on the horizon and in another moment it was out of sight. The fellows off watch stood silently, watching for the geysers to shoot up over the horizon that would tell of depth bombs. Then, Cookie, a Chinese who likes to wear his derby hat all the time and hangs it on a hook only when taking a shower, sounded chow time and the fellows all hotfooted it for the mess, their appetites plain on their faces.

Being a landlubber at sea at such a time is a nervous business because the Navy kids on the armed guard and the Merchant sailors like nothing better than to put a needle in you. You learn very fast that the great point about a *contretemps* is to be where it ain't, that if you're where the torpedo lands you haven't got even the chance of a monkey in a dream by Frank Buck, and the only time you have a chance is if you're aft when the torpedo strikes forward or forward when the torpedo strikes aft. Then the gunners and the sailors get into arguments as to where the subs like to hit first — the steering engine in the stern, or the engine room 'midships or the bridge and control room forward.

5

You get to listening nervously to the arguments. You know there's nothing you can do and you just have to trust to luck, but you are tempted to press your luck anyway, especially when you have nothing else to do. But when you stay aft, you think of one thing and go to the waist of the ship and think of another thing and move forward, and what do you see there but the bridge, sitting up there like a tin duck in a gallery.

I think I covered the ship gingerly from bow to stern seven or eight times before I wound up on the bridge with the third mate. There was a hearty steam and clatter coming up from the officers' mess below. The ship's carpenter came clambering down from the flying bridge, holding his head where he had smacked it against a turnbuckle.

'What's the matter, Chips?' asked the mate. Chips grinned and took off his dirty white cap, showing a small gash on his forehead.

'The torpedo hit me here,' he said, tapping the place and smiling.

The boatswain's mate in charge of the armed guard yawned. 'It looks like it might be a long night tonight,' he said. 'I think I'll hit my sack and get ready for it.'

Some ordinaries and a few A.B.'s were out chipping and painting. 'Don't fall off,' the captain called to them jocularly. 'If you fall off I don't stop for you, not for nothing.' He spoke with the heavy contentment of a large man who has just eaten his fill.

The escort vessel that had disappeared was still ranging around out of sight. I kept thinking of subs going along in step with us under the sea, waiting for twilight or moonlight to close in and kill. 'Everybody is so damned casual around here,' I said to the mate. 'You'd think they were a bunch of limeys in some movie with Leslie Howard.'

'It's always like that,' the mate said. 'I made a few trips to Murmansk and once, when we got just fifteen miles of open water between the Germans and the ice, we pass two lifeboats full up with guys. We offer to take them aboard. "Hell, no," they say. "We got ours once and we'll stay where we are. You're going to get yours soon." In the four and a half hours we had to wait before getting ours, the ship was just like this one is — everybody minding his own business.'

I shook my head disbelievingly.

'Hell,' the mate growled. 'You're so sure we can't lose the war. Everybody is so sure. How can you be so sure unless there's American guys all over everywhere who don't get all girly and goosey every time there's a chance of their being killed?'

Still I didn't believe that a twenty-five-year-old, even if he is a boatswain's mate, first class, could prepare himself for a night of battle by going to sleep. But when I passed his sack, there he was, all stretched out, as deep in sleep as a baby, his chubby face all soft and pink with sleep.

The morning alarm, the convoy lurching off course, the escort going baying down the horizon and so forth, had been like the dropping of the first shoe. Now it was a question of lying still with eyes wide open waiting for the second shoe to drop.

The afternoon wore on slowly. I spent some of it picking up sunburn on the poop deck aft, along with a wiper out of the black gang who had been in the advertising business in Milwaukee making a hundred and forty dollars a week when Pearl Harbor batted him in the belfry. A gonzil, as they call the gunners, was taking clips of bullets out of ready-boxes and spreading them on deck and cleaning them with a stiff brush. He was a nineteen-year-old farm boy out of Missouri, and as he worked with pursed lips there spread out from him the slow, patient, drowzy atmosphere of chore time down on the farm.

'I figured the Merchant Marine was the more important of the two, anyway,' the ex-advertising man said. 'I couldn't get into the Navy on account of my eyes, and I couldn't get a deck job here on account of my eyes won't let me be any good on lookout. So I'm down in the engine room, nervous there, let me tell you, boy, plenty nervous there, because who gets killed if not the engine room crew? You can see it yourself. When there's trouble, who has the dirty, dangerous job of turning his back and trying to run away? Not the Navy, no; us, the Merchant Marine.'

I asked the gonzil if he was brushing off the bullets to keep the Japs from getting blood-poisoning.

'Blood-poisoning will be the least of their troubles when these hit,' he said.

'Tonight?' I asked.

'Well, you can't tell about the Japs,' he said. 'They throw rocks any time. They're crazy for throwing rocks when you don't expect them.'

After that I moved to 'midships near the lifeboats and watched a bald-headed ordinary seaman who, two months ago, was wearing a boutonnière in his lapel as assistant manager of one of the large hotels in San Francisco. He was a curious sight as he assiduously tangled up his eleven-dollar pointed shoes and affable, deferential manner in stubborn deck gear. The afternoon wore on and wore on and then wore out, and general quarters sounded and I went to the bridge with my pencil, prepared to go down with my pencil.

The twilight was remarkably beautiful. I did not pay much attention to it. The sun was finishing up with a purple passage, but I kept watching the troopship on ahead and wondering what was going on among those fellows. The Merchant Marine and the Navy had stacked up very nicely, showing plenty of guts in the emergency. But this was their war out here. They were the fellows fighting it, while the troops were just being carried out to their war and in the meantime had nothing to do but sit and watch and feel all sort of exposed, like pins in a bowling alley waiting for the ball to roll.

I had ridden across the country to this convoy with some of those troops and it hadn't seemed to me that the country had done anything to build them up for the experience they were now going through. When the old A.E.F. shipped over twenty-five years ago, they sang it out, and singing makes a man feel like a flag flapping in his own wind. But the new A.E.F., if it sang, would have to sing like this:

> Where do we go from here, boys?
> Where do we go from here?
> Anywhere from (restricted information)
> To a (restricted information) pier.

10

So the new A.E.F. does not sing.

A lot of us took off for here from the Pennsylvania Station in New York. It was nearly midnight. The station was gloomy on account of the dimout. There was a big crowd pushing around in it, but everybody was quiet and it was easy to pick out the fellows who were there to say, 'Good-bye, mama; hello, war,' and easy to pick out their girls. They weren't fooling when they kissed, but were hungry and earnest. When they put their faces together, it was a moment like a bugle blowing, all solemn and with the whole heart standing at attention.

'Come back, boy,' you could hear one side saying with their silence, and the other side saying silently, 'I'll come back all right, I think.'

Then the train started to pull out and girls and women and fathers walked along the platform, some smiling a little, but most just trying to look brave, walking faster and faster with the train while the soldier boys stood in the vestibules or crouched in their berths, faces pressed against the glass, looking out until the lights ended and the tunnel began and there was nothing to look at.

This wasn't a troop train. The troop train was on ahead and this was some kind of a mixed-up something for the left-overs and casualty replacements. A lot of the fellows felt restless and lonely

11

and headed for the club car to get some pain-killer. But the space there was all taken up by business men and so forth. The business men hadn't had anybody to look out the window after until the last minute and were experienced travelers, so they had beat it for the club car fast. They had got all the seats there were and the comfortable standing room, too, and were drinking nightcaps and bellyaching about this and that, mostly the new taxes.

A lot of the soldiers stood around, first on one foot and then the other, listening to the bellyaching and hoping some windbag would get up and make room. But more of them shacked up in the little smoking-rooms, sitting on suitcases and duffle bags and washbasins, not talking, just listening to whatever was said. They were afraid to talk because this was not a troop train, not one of those gorgeous 'Tokyo or bust' trains that nobody ever seems to get a ride on, and everybody was supposed to keep his destination secret from everybody else.

That gray good-bye feeling lasted all through the country, and the beer never did get to taste really fine and mellow until Nebraska. Nebraska was the white spot. It was the only state between New York and California that spent any time gingering up kids who were sitting lonely in a crowd and wondering if they'd ever live long enough to become twenty-five years old. All the other states stayed home, feeling shy, no doubt, about making a fuss over some guy when they didn't know where he was going, maybe home on a pass or to Ruby, Arkansas, to guard a power station. But in Nebraska they didn't care a hoot where a kid was going so long as he was on his way. In all the little towns there, they had women and girls in bright dresses and in their best church social manners running up and down alongside the trains with baskets of fruit, throwing oranges at every uniform they saw, and friendliness and a feeling of excitement. It lifted up the whole train.

But by night the train was out of Nebraska and into that gray good-bye feeling again. Going on the ship was like going on the train, except there was nobody to say good-bye to. There was no singing or laughing, just the businesslike clop-clop of G.I. shoes clumping up the gangplank and, when the ship let go its lines, it gave a sallow little hoot, a kind of hoot-in-hell thing, and went slopping and splashing out into the stream and that was all there was to it. This is a story that is going to be the first chapter of half the novels of the next twenty-five years, but that's all there was to it. Not like the old days at all when a soldier got enough razzle-dazzle on his way to the ship to carry him through

to where the nervousness could go out of him because he was given work to do.

A few hours out of port, the little tell-tale splashes
could be seen along the sides, telling of fellows being
seasick. In a ship as crowded as that, when one
gets sick everybody does. And altogether, all the
way down the line from Penn Station to midocean,
it seemed to me an especially punk way to start a
bunch of guys off for a fight. Necessary, true, but
punk just the same, particularly so now when we
were all standing around waiting for the Japs to
drop the other shoe.

Nothing happened at twilight and then it was
dark, real black dark. The captain said he liked
black nights, the blacker the better. It gave him
the feeling of cuddling up snug under a blanket,
he said. He is a man who likes to sleep with a
blanket pulled right over his eyes.

But the blackness did not last long. The moon
poked up over the horizon and was so big and
swollen it looked blistered all over. There wasn't a
cloud to hide it. It looked like one of those moons
on a Pacific cruise poster and the captain swore at
it with real bad language.

'Nice shooting weather,' I said, and the captain
stamped sullenly up and down the bridge.

About this time the wind changed. I had been

14

watching the silent black hulk up ahead that was the troopship without being able to make out anything on it except what seemed like huge, thick clusters of white-petaled flowers, which was how the troops looked standing on the open deck with their life jackets on. But when the wind changed, I heard them singing. The wind brought it back in puffs and gusts, fine harmony with hundreds of big, strong voices bellowing in it. It sounded like kids on a hay ride and like what the President said, fighting on 'An extension of Main Street.' For a while I couldn't make out what the song was, but in the middle of the second chorus discovered that those crazy kids up there were singing 'Shine on, harvest moon,' as their answer to the needling from the sailors.

The Japs never did drop the second shoe, so nothing really happened. The escort vessels pushed them down or scared them away or maybe even, down there below the horizon, knocked them off. But in the meantime this generation of kids, the so-called 'soft' generation, 'softened up,' according to experts, by this and that, WPA, boondoggling, youth administrations, automobiles, labor-saving devices, and so forth, had a chance to prove themselves.

The armed guard, which is regular navy, the

Merchant Marine and the troops — kids for the most part — stood very well the little harrowing-up they had to go through. A man doesn't have to stick much more than his big toe into the war to find out that anybody who calls the present generation 'soft' is just whooping in an empty barrel.

CHAPTER 2

Sky Road and Sea Road

★

CHAPTER 2

★

Sky Road and Sea Road

IN A UNITED STATES ARMY FLYING FORTRESS EN ROUTE TO THE SOLOMON ISLANDS

Even going up to the line is different in this war.

There were some Army girls and Army wives standing around the plane, back there in Oahu, Hawaii, a lot of blue water ago. The plane had five Jap flags painted on it, meaning five Zeros made naught, and had guns sticking out all over it like quills on a poked-up porcupine, but with the girls there it didn't seem too much like business.

The girls' bright, pretty dresses were blown out stiff behind them by the wind. Their hair was blowing backward, too. They were giggling and flossing around and the whole thing, with that blue Hawaiian sky over it and the thick yellow sunlight coloring it, didn't look like going up to the front line at all, but like something in a magazine. There were even two low-hung roadsters in the background to complete the picture.

'I don't know why they won't let me go back with

you,' a fellow said in a low voice. 'I feel all right.'
This was Assistant Radio Man Kendall Shoop of
Harrisburg, Pennsylvania, who had come down with
malaria in the Solomons and was still technically
hospitalized.

'You'll be back soon enough,' the pilot, First
Lieut. Ed Loberg of Tigerton, Wisconsin, told him.

'I'd like like hell to go back with you,' pleaded
Shoop, and Loberg said he knew how Shoop felt but
to let the doctors be boss.

There was a whole silvery chorus of soprano good-
byes as we climbed into the plane and suddenly one
of the Army wives there, a real pretty-looking
young girl, cried out, 'Bring back my husband!'
Everybody smiled at her encouragingly. 'Safe,' she
said. I looked at her closely for the first time. Her
whole face seemed drawn into a point and was gray
and sick-looking.

Her husband, a bomber pilot in the same squadron as these boys, had been rumored down somewhere in the Pacific after a tussle with warships off
the Solomons. Then he had been officially reported
missing in action. That was late last week. Then,
a few days ago, he had been rumored picked up, and
last night still another rumor reached Honolulu that
it wasn't he who had been picked up but some Navy
fellows.

'The plane'll get up fast,' the co-pilot, Lieutenant B. K. Thurston of Indianapolis, said. The Fort wound up, lumbered, ran, faster and faster, on its toes now, and then was in the air. It swung around and came back to 'buzz' the field. 'Buzzing' a field consists of paying your respects to it by slamming across it so fast and so low that the wind you make takes paint off the hangars. The girls were still standing around the roadsters, all waving and laughing.

The fellows buzzed a lot of other things, but by the time I dared to open my eyes again we were safely out over the Pacific. It seems funny to call that 'safe,' but that's the way it felt. The enemy has got one of his big teeth stuck deep into us there, and what we were doing was skirting the edge of that tooth to get where we could do some drilling on it. That's a real old-time bucket of blood that we are heading for in the Solomons. There's no fun for anybody in the Solomons and no fun getting there, but it felt safe all the same because there was nothing to buzz on the water and the Fort itself is the most comfortable plane I've ever ridden, more comfortable even than that luxury liner, The Clipper to Europe.

A portion of the hop included an over-water flight that is longer than the one that made Charles A.

Lindbergh famous. Lindbergh had Europe to aim at. Our navigator, First Lieutenant Robert D. Spitzer of Anderson, Indiana, was aiming at something no bigger than a dime lying flat in the middle of a watered-over wasteland. However, he didn't seem worried.

'If we miss and run out of gas, we can sit down on the water and have about thirty seconds to get out,' he explained. 'There's a whole technique to it. We all, except the pilot and co-pilot, stand back in the radio compartment. Then, when we hit, Sergeant Holbert [Radio Operator George R. Holbert of Lamar, Colorado] throws the rafts out of the escape hatch and we all pile out after them and Loberg and Thurston come running like hell after us.'

Then the engineer, Paul A. Butterbaugh of Altoona, Pennsylvania, discovered there were not enough rafts to go around. The rafts could carry ten and there were eleven of us on board. 'Oh, that's all right,' said Co-Pilot Thurston. 'The pilot and co-pilot are always killed in these water landings anyway.' Everybody laughed and looked along the narrow catwalk down which Thurston and Loberg would have to make time if we sat down. It was narrow enough to be cleaned with dental floss.

The weather included one 'thunderhead,' as they call the massive, beetling cloud formation that many believe killed Amelia Earhart in this part of the world. The giant plane bounced and bobbed and fluttered like a bird in a hot Badminton game and, when it was over, Thurston, who had been at the controls, was mopping his face. 'I kept wondering if I had remembered to bring my reflexes along,' he said, and I smiled uneasily at his grin.

At the end we knew we were getting somewhere because some P-shooters came out to buzz us. P-shooters is what the fellows call our pursuit planes. They lined up just out of range and gunner Ellsworth W. Jung of Milwaukee, Crew Chief Harry B. Brand of Stanford, Illinois, and Assistant Engineer Everett W. Gustafson of Malden, Illinois, all grabbed guns and started chattering at the P-shooters with their teeth.

'Now you'll see what an attack is like,' said Lieutenant Spitzer. He was hopping around excitedly in the nose, looking out first one window and then the next, studying the technique of our boys and comparing it with the Japs'.

'That's the way the Zeros line up,' he said. 'Single file, just out of range. They ride along like that, scared to death, trying to hop themselves up because they know they're going to get it when they come in. Then you see one peel off and it's just as if he said, "What the blazes, nobody lives forever."

23

He peels off with a kind of shrug and comes at us head-on and we pop him, and as he goes down you can see the others strung out, out there, bobbing and dipping a little like they were shuddering.'

The P-shooters came at us head-on and from the sides, diving and looping. 'No good!' Mitchell shouted. 'That's it!' Spitzer shouted. 'That's the one! That's git or git got.'

The little toylike planes slashed at our wings and made passes at our belly and banged at our nose, and the Flying Fort floated on steadily and serenely like some giant, bull-shouldered animal disdaining to shake off its flies.

That's the way the Fort acts in a real attack, too, the men say.

These men I am traveling with have been through assorted species of hell for the last three months, flying combat one hour in every seven. They went to Hickam Field, Oahu, on a repair job and are now flying back to rejoin their squadron in that pig-sticking and blasting job going on in the neighborhood of Guadalcanal.

When the story of this sky and sea road is told, after the war has been won, the chances are pretty good it will come out as one of the most extraordinary achievements in a very extraordinary war indeed. The part of it I saw was strictly an Army job, but I am told that other parts of it are feathers in the caps of the Marines and the Navy. The road was punched through under forced draft by men heedless as ants working on the edge of the vast catastrophe that smote the Eastern Pacific and is still perking there bloodily. It's a 'this far and no further' line drawn around the well-fanged Jap where he has bit into the toothsome Pacific, and it is hoped the line will one day turn into a strangler's cord around his neck.

The part of it we see as we sit down at night and hop off at dawn consists of the kind of islands which, much to the bewilderment of the fellows now there, people used to throw off their wives for and dream about in the old days of, say, 1938. Some of the islands look like those in the comical cartoons about castaways. Others are populated sparsely by peculiar folks whose cannibalism and head-hunting antics and scanty hip-wear form the current legends of the new A.E.F. A vast crawl of traffic buzzes this way and that way, making some of the places look a little like a beach resort road on a peacetime holiday. The fellows on the islands are something like traffic cops at Reuben's Junction. But, whatever the traffic, the islands still remain that kind the cartoonists draw with a black line around white space — bleached and shrunken rocks, a mile or a few miles long and less than that wide, unequipped with anybody's favorite movie star or anybody's ten favorite books.

The uniformed 'castaways' stationed on these islands have one of the rugged jobs of the war. They know that, if the Japs want to get anywhere — and it seems to be a habit of theirs to try — they will have to hit them first. When such a time comes, as, for instance, it came to the Japs who had the same job on Makin Island, then these fellows will have to drop their job of directing traffic and take what the Japs give and hold on with whatever they've got until help comes from maybe as far as a thousand miles away.

But that's only part of their troubles. Another part is what may be described as an 'interesting sociological experiment.' This is the first time ever that large bodies of men have been snatched up out of civilization and dropped on rocks in the middle of a watered-over wasteland to live and work there, cut off from the world, in tropical heat and in a loneliness that cannot be imagined.

From what I was able to see, during blacked-out nights and dawns that lay in the sky like aching hangovers of the night before, the uniformed 'castaways' seem to be doing all right in their 'interesting sociological experiment.' In one of the caravansaries en route was a hilarious reference to the fact that pillows are scarce on the rim of this bucket of blood known hereabouts as 'the Solomons deal.' A sign in the bunkhouse reads: 'Do not take pillows and slipcovers with you when you leave. This is not a hotel.' And on another island, when I was wandering around looking for one rock that stuck enough up over another to be photographed, a private interrupted to remark courteously, 'The tree, sir, is over here.' His manner was that of the butler in the *New Yorker* cartoon preparing to draw the curtains for the night and saying, 'Have you finished with the moon, madame?'

On islands where there are natives, the fellows have a better time of it. They do what the natives do in the way of handicraft with palm leaves and shells and this and that, and it seems to be true that, wherever Americans are, the 'natives' like them. There is not going to be any repetition of British experiences in Burma on these islands, because the American boys squat right down with the people and teach them what they know and learn what the natives know.

There is a native colony on one of the islands made up of men and a few of their wives who beat it out of the Gilbert Islands in canoes because the Japs were paying them off for their work in funny-looking money. A sergeant, whose name I dare not mention because his family must not know where he is, that being a military secret, took me down to see them. 'Seet don!' the brown-skinned boys shouted when I came in, and I found there was no room to sit because the soldiers had taken up all the space. They were exchanging information in a weird pidgin English about the best way of stringing shells to make a necklace, and the way to weave cloth out of grass. After a while, somebody produced a ukulele and one of the blacks produced a love song out of that apologetic instrument that had everybody thinking of his girl back in Brooklyn or Dayton, Ohio.

One of the reasons why this remarkable route must be written about thus vaguely and inconclusively is the camouflage job Army engineers have done along it. Nothing is what it seems, even from heights of a few hundred feet. Runways have been made practically invisible from any place. Hard-surfaced roads do not exist for the human eye perched aloft unless a truck comes along to point them out. As for hangars, depots, barracks, entrenchments, bunkers, and clubhouses, you have to fall over them to find them, even in daylight.

Most of the men traveling this route are ferry command pilots, rugged boys who fly around the world the way gay blades in Lincoln, Nebraska, used to go up to Omaha for a dash of fun of a Saturday evening. They don't see much of what they go through, and what I remember most clearly about all of those I have bumped into and bunked with in the last few days is one lad who had the cot next to mine at a way station in the New Hebrides. There had been a mix-up and I had to dig him out of my cot at nine o'clock in the evening. He got up with eyes closed and fell into the next cot and slept there undisturbed until 4 A.M., when he was wakened for the take-off. He stumbled swearing into the blacked-out night and there was quite a crash as he fell over a snazzy rocker captured by some officer somewhere and left right outside the door to catch the afternoon sun.

'Oh, jeepers!' he cried in pain — and 'jeepers' was exactly what he said — 'I hope this war doesn't last four years like they say.'

Anybody who has ever been to the movies has had several good times crying over a scene where dashing young aviators descend on a smacked-up buddy in a hospital and cheer him up in a way that is rough and manly, but nevertheless shows the hearts of gold beating tenderly underneath. Well, this is the way such a scene works out in real life.

The wounded pilot was Lieutenant James Lancaster of Temple, Texas, and his buddies were the officers of the crew with whom I am flying to the Solomons — Pilot Lieutenant Ed Loberg of Tigerton, Wisconsin, Co-Pilot Lieutenant B. Thurston of Indianapolis, Navigator Lieutenant R. D. Spitzer of Anderson, Indiana, and Bombardier Lieutenant Robert A. Mitchell of Washington, D.C. They found out by accident at one of those islands around here that 'Lank,' as they call him, was among some beautiful nurses near by, hopping around on crutches and hollering to get back with his squadron. 'Ah doan' lak this here pee-yure life,' he had been hollering to all the nurses. 'Let me back in the war.' And all the nurses thought he was cute.

'Did you hear about Si?' Lank asked, referring to Lieutenant Sidney I. Dardan of Waco, Texas. Nobody had heard. 'Well, he sat down on the water a day after I got mine and was in here to tell me about it. They had two or three minutes before the plane sank and all nine of them stood there by the radio hatch throwing things into the raft. They had only the one raft. The other one was folded up somewhere in the muck below.

'They kept pitching. Boy, how they pitched! Chocolate, peanut bars, ham sandwiches, fruit, pillows, blankets, enough to make that raft some kind of a damned yacht or something. Real cushy. Then they pile into the raft themselves and what do they find but they are going straight down. The raft has a big hole in the bottom of it and all the stuff they've been throwing into it has just kept right on going straight to the bottom. So there they are, not in a yacht at all, but riding the Pacific bareback.'

Lancaster laughed uproariously and everybody here laughed until the tears ran out. Then somebody said, 'I guess it wasn't very funny for Si.'

'I guess not,' said Lancaster and laughed again, thinking of those nine surprised faces in the dark there. Then he quieted down. 'There were nine of them in this one raft, made for five midgets on a diet. They had no place to sit and nothing to eat or drink or anything and were in it for seven days, before being picked up by the Navy. Sharks followed them all the time and they'd shoot every one that tried to come up through the hole in the bottom.

'One of the fellows died of exposure.

'The Navy ship that picked the boys up had a real ice-cream bar on it.'

'Pretty soft,' somebody said, and somebody else demanded of Lank what he meant by getting shot and letting down his public that way. Lank wanted to know what public. When he heard that the newspapers and radio had made a hero of him for shooting down a four-motored Japanese flying boat in three minutes, he kept saying, 'Oh, that; oh, shucks, that one,' and blushing with pleasure. Then he told how he had got his up Bougainville way.

'We ran into some Zeros,' said Lancaster. 'One was high up, setting himself for a dive, and the boys at the guns poured it on him. It was the prettiest sight I ever saw, those tracers going into him like red pepper into an egg. He started to whup from side to side — you know that seesaw they get on before falling — and I thought, "Well, pal, good-bye," and swung in under him to get on with the bombing when whop! Something like a horse kicked me in the leg.

'It was just a little old bitty .707 (about the size of our thirty-caliber bullets), but it knocked my foot off the controls and I felt awful foolish because I just couldn't get my foot back where it belonged. Well, that was all there was to it.'

This morning we said good-bye to Lank in a way the movies would approve. We buzzed past the hospital real low twice. Lank, a tall, whippy-looking lad, was standing out front watching us, and as we passed he lifted one crutch straight up into the air to salute us.

This is being written on a case of hand grenades on a jungle airport on the edge of the business. The fellows with whom I have been going up to the line have got there. Tomorrow they go to work. In the meantime, one of them is reading two-month-old funnies. Another is playing poker. Others are shooting the breeze with the rest of the squadron, one of whom — Lieutenant David R. Flittie of Brookings, South Dakota — has just shouted, 'Why any tourist in his right mind ever wanted to pay thousands of dollars to see this island beats me.'

It's just before the battle, mother. But there again, it's not like the movies put it. The flies and malarial mosquitoes are so thick you can brush them off your arms with a broom.

CHAPTER 3

The Loss of the Wasp

★

CHAPTER 3

★

The Loss of the Wasp

A SOUTH PACIFIC BASE IN THE GUADALCANAL SECTOR

The aircraft carrier *Wasp* had bad luck and in waters outside of here ran head-on into one of those rotten twists of happenstance that turn up during war. About three o'clock on a clear, tranquil, sunlit afternoon, it took three torpedoes from a submarine. It was built to take three times that many without blanching, but the first one nosed into a bomb that blew blazes all over everywhere. The second tin fish swam clear through. After that the third one did not matter. It was just a kick in the head of the dying body of the *Wasp*, helpless to staunch its wounds.

The *Wasp* died a lingering death, burning for hours with vast, blazing spewings that looked like the mouth of war itself opened wide. Our ships hovered around it as anxiously as a covey of doctors around an agonizing millionaire's bedside, but in the end, United States destroyers reluctantly had to give the *coup de grâce* with torpedoes.

The loss of life on the *Wasp* was comparatively light, considering the way its luck was running, and it was reduced by the heroism of the men aboard; heroism which, judging by the conduct of the Navy in this greatest and most prolonged naval battle in history, seems to be as much a part of the standard equipment of our fellows as the dog tags they wear around their necks. A substantial ingredient of heroism is the ability to keep thinking in the face of disaster and to conquer the instinct of self-preservation and never to let go of anything until the experts command otherwise. A *Wasp* survivor story that best illustrates this feature of the United States Navy is told by Ensign John Jenks Mitchell, twenty-two-year-old Annapolis graduate from Washington, D.C. As his shipmates put it, Mitchell established a new world's record for an involuntary high jump by getting himself blown thirty feet high and sixty feet away. The record for a survivor in the last war was believed to have been thirty feet high and a few feet away.

Ensign Mitchell is now convalescing very nicely here from a broken leg. His major distress seems to be that the broken leg temporarily is preventing him from passing the physical examination necessary before receiving a promotion to lieutenant (j.g.) which he has been given.

Mitchell described his experience with considerable unilateral amusement to an array of shocked faces, one of them belonging to his brother, Second Lieutenant Robert A. Mitchell, a bombardier on a Flying Fortress which figured in the battle of Midway, softened up the path for the Marines at Guadalcanal, and is now working over Bougainville. The other faces belonged to Lieutenant Mitchell's plane mates.

'I am somewhat ashamed of my story,' said Ensign Mitchell. 'As far as I am concerned, it consisted of a loud, unruly noise — something like a railroad train going up a flight of stairs — and the next thing I knew it was ten days later and I wanted a cigarette. My friends here have told me what I did and saw during those ten days. In fact, this ward has been something like a fraternity house on the morning after the night before, each man telling the other what he did while under the influence of — say, high explosives.

'General quarters had been on all day and I had just been relieved when a seaman in the crew of my gun station said in a somewhat unseamanlike way, "Mr. Mitchell, what's 'at funny-looking thing out there?"

'It was a torpedo wake. I sounded an alarm in a voice that may be described as similar to a thrilled tenor, for the torpedo was coming in right under my feet. I remember that the *Wasp* started to turn, but being a careful sort of man, the next thing that happened I was going hastily toward the bridge. The torpedo influenced my decision. It came in under my feet apparently, although my vision may have been prejudiced, close enough to peel the soles off my shoes. It missed my shoes but hit the bomb I was standing over.

'They tell me I landed right at the feet of my superior officers on the bridge in a posture unbecoming even an ensign, being flat on my back.'

Ensign Mitchell laughed heartily and Lieutenant Mitchell, who has been in several battles against the Japs with his kid brother, but was seeing him now in a quiet, clean little officers' ward near here for the first time in two years, joined uncertainly in the merriment.

The bridge, it developed, was from twenty-five to thirty-five feet above and exactly sixty feet away from where Ensign Mitchell had been standing when he was blown up. In disclosing this fact, Ensign Mitchell laughed uproariously again and said, 'In fact, I am thinking of putting in my chit to qualify for landings on the flight deck.' Then becoming serious, he added in a low tone: 'I guess God had his hand on my shoulder that time. I started for the pearly gates but he stopped me in midair.' As far as can be learned here, Ensign Mitchell is the sole survivor of his gun crew.

After that, another remarkable thing happened to keep Ensign Mitchell alive. In the condition in which he landed there was considerable evidence to indicate that he was dead. But even in the awful circumstances then shaping up over the whole ship, the Navy refused to let one of their own be marked off for the Japs until the experts called the score. The experts were far from the bridge at the time, so Lieutenant Courtney Shands, who is now a few beds away and looking in very good shape, and Commander Beakley strapped the ensign to a stretcher. Then Lieutenant Shands dropped down to the explosive murk below where airplanes and their ammunition were popping like balloons on New Year's Eve and dug a raft from an airplane for the unconscious boy. The ensign was lowered carefully to the calm sea. But the raft was topheavy and it turned over, leaving its occupant tightly secured under some half-dozen feet of water. The sea was full of sharks which were attracted by the bloody gruel of bodies, and everybody in it was under great temptation to make tracks away from there. But Lieutenant Robert Slye struggled with the raft until he righted it, pausing every now and then to thrash at the sharks with his feet.

CHAPTER 4

Slugging It Out

★

CHAPTER 4

★

Slugging It Out

GUADALCANAL, SOLOMON ISLANDS

This is no banana war going on in the Solomons, involving potting hopped-up Japanese killers from trees, but the fightingest engagement involving American troops since Bataan.

The reason for the necessity of victory after victory for the Americans holding a spot of land on one-hundred-mile-long Guadalcanal is that the Japs keep on sneaking men and guns into remote places, accumulate a striking force in the jungles, and then hit with it.

The Japs were pretty well convinced by their parade down the Pacific that islands were impossible militarily to hold. Our strategists, trying a stepping-stone route to Tokyo, are out to convince the Japanese that islands are impossible militarily to hold only for them.

The prolonged Japanese effort to break off our toehold in the Solomons is continuing today in a now tragically familiar tempo. The Japs are paying dearly for their sudden access of ambition, but presenting a bill and collecting it is no picnic.

Twenty-two Japanese high-level bombers overtured the present show with thousand-pound crumps, starting a few minutes after noon yesterday, October 13. Marine anti-aircraft gunners laid a perfect pattern under this flight which would have knocked down every plane in it except that the shells blew up exasperatingly just two thousand feet under the targets.

A second wave of Jap bombers came over two hours later. At least three of these came down in smoke and burst against the earth like Roman candles.

About eleven o'clock last night the enemy began a concentrated bombardment which the Marines, many of them veterans of Pearl Harbor, declare was the heaviest bombardment of the war for them. It was apparently conducted by Jap battleships with blinky-eyed gunners, since fourteen- and sixteen-inch guns threw salvo after salvo toward Henderson Airport without doing much more than dig excellent foxholes for future use in the mellow black earth around it.

The bombardment was lifted shortly before dawn and tension relaxed in our foxholes when — wham! — Japanese field artillery opened up on our positions from near-by hills. This is what is happening now. Our artillery is barking back throatily, pausing only when our dive-bombers get over the target.

We are using mostly pursuit planes for dive-bombing. The planes take off in the face of very accurate fire, skittering along the runway with dust kicking at the wheels and giving them an extra lift into the air. The pilots find the Jap batteries by aiming their noses into the shells spewing from them. A particularly annoying battery was silenced by a pursuit pilot who came in with a screaming dive, found the enemy's fire missing him, climbed steeply thousands of feet, then turned and slammed himself down directly into the mouths of the guns, from which, a flicker later, a deathlike silence ensued.

For obscure reasons known only to Fate, this plane escaped damage and was cheered to a safe landing. Every man on Guadalcanal was ready to give this pilot a medal, or, what's more useful, a glass of cold water.

While this was going on, our ground troops were girding up the bloody beaches and hills, as a Jap naval task force, including at least five transports, was reported nearing the island. Army bombers are hunting them now, loaded with a foretaste of what is awaiting them if they get through.

Jap bombers are not accompanied by Zero fighters, indicating that a heavy toll has been taken by our Flying Fortresses and by Army, Navy, and Marine fighters. Also, there have been no planes yet from Japanese carriers. Exactly how many Jap flat-tops have been scratched is better known on the mainland than here, but it seems to be plenty.

The Japs added wrinkles of their own to the German-style *Blitz* by inserting snipers amid the crumps. The snipers made the tactical error of trying to plug Marines who were enjoying a swim in the sea while the first bombs were falling, and compounded the error by firing into the mess-hall while the troops were trying to enjoy lunch. Incidentally, trying to enjoy lunch involves some trying, since every time you open your mouth it fills up with flies and malarial mosquitoes.

The Marines regarded as unpardonable this intrusion of bullets into the moments of tropical languor they had snatched from battle. The snipers came in under their own bombs and the Marines went stalking for them under the same bombs, fighting a stealthy, itchy-nerved private war under a canopy of bursting high explosives. The Jap snipers were snipped off.

During the night the fighting guts of our men were brightly illuminated by the glare of flares and bursting shells in a genuinely nasty bombardment.

Our men, crouching face down in foxholes, heard a boom-boom first, then silence, then felt the earth tremble all around. The earth shook like a hammered gong and then the black sky was filled with wild, wailing shells. After that came a vast blast and frequently the snakelike slither of shrapnel plowing huge furrows in the earth. Pieces of shrapnel weighing more than eighteen pounds were found in the fields this morning.

This terrible, passionate symphony of war went on for more than two hundred and forty minutes, each one brain-smothering, but the dawn's light revealed unbelievably few casualties. Each man had his own story to tell of a narrow escape. Some men did not have time to reach foxholes and lay between logs, under leaves, or tried to dig into the earth with their own noses and chests.

Three enlisted men in a bomber crew lay face down on the earth with arms around each other. When a shell burst directly overhead and the awful slither of shrapnel ceased, all three laughed and agreed a miss is as good as a mile. A moment later the man in the middle said in a faint voice: 'I think I am hit. I cannot feel my back.' Shrapnel had broken his back without touching the men on either side of him.

A Signal Corps man at a switchboard running to

the front-line trenches remained at his post throughout and, evidently to relieve his nerves, began broadcasting by telephone a blow-by-blow description of the bombardment to the fellows lying in. He could see the muzzle-flashes from where he was located. 'Here comes a good one!' he would cry. 'These are short.' Suddenly his voice became as ecstatic as if a blonde mermaid had come waltzing from the sea. 'Boy, here comes a peach!' he exclaimed. Which was a bit too thick for the fellows out in it. A whole chorus of profane baritone bawlings rose from all over everywhere in the seething blackness. The general tenor of the remarks was expressed pithily in what sounded like very good Bostonese: 'Shut that man off! He's making so much noise I can't sleep.'

Between bursts an American pilot reported with great amusement a conversation he had had in the morning with a captured Jap bomber pilot who claimed to be a graduate of Ohio State University. The Jap, he said, seemed puzzled and remarked, 'I understand what we are fighting for — Togo — and what the Germans are fighting for — Hitler — but your Marines seem to be fighting for souvenirs!'

During the bombardment the same Jap observation plane which has been bombing the airport every night for months came over to spot for the artillery. 'Here comes Washing-Machine Charlie!' cried the Air Corps men. They call him that because they have an ear for motors, his being a peculiarly grinding, snorting, persnickety one. But the Marines, who have an eye only for character, cried, 'Here comes Lousy Louie!' All seemed glad to have at least one old friend join them in the soup.

Artillery from ships and on land and aerial bombs called the tune of the battle of the Solomons in day and night fighting, October 14 and 15, with the Japanese and American forces slugging it out toe to toe.

A Jap naval task force battered through into the mouth of our positions, leaving a long, crimson trail of dead and dying men and burning, listing, and foundering ships stretched out for miles to sea. Japanese troops scuttled up the beach under a blasting barrage and disappeared into the trees, where they are now trying to form under fire from United States Marines and Army infantry.

The Army infantrymen are getting their first taste of hot steel and seem anxious to prove that they are as tough as the Marines. A measure of their mettle is the fact that some two hundred stowaways are reported to have stuffed themselves into nooks and crannies aboard the Army transports, technically deserting their companies in order to hook up with outfits going into action. These men do not know whether medals or courts martial face them and are not particularly concerned about that now, that being a remote bridge to cross.

Up to now the land forces were not yet heavily engaged but were merely sitting and taking it and biding their time. The battle is rapidly mounting to a climax and they know their time is coming soon. A decision ought to be reached in a few days.

Yesterday's work was done mostly by our Navy and Marine flyers and by Army Flying Fortresses under Colonel 'Blondy' Saunders, a former West Point football star. The Fortress pilots reported the heaviest anti-aircraft fire yet, with land-based Zero fighters, which they presumed to be from a carrier or carriers, sashaying and fandangoing and throwing lead at them up to the edge of the flak, then running around the flak to pick them up when they had finished their bomb runs.

The bomb runs lasted on an average a minute and fifteen seconds, which is a very long time indeed to float steadily on an even keel with every appearance of serenity in a sky which is literally blacked out by anti-aircraft shells. The pilots fought the blasts with reflexes to keep their huge planes steady, and emerged from their runs drenched with perspiration and breathing heavily from exhaustion.

Direct hits were reported on two transports and a battleship by one squadron alone. All the Fortresses returned and reported the Zeros were 'very fresh today' and getting their cheekiness slapped good. All the Fortresses had bullet holes in them, but the point is that all returned, and the ground crews worked all night long, with guns at the ready for land and air invaders, to have the planes back in the air today.

The ground forces on Guadalcanal are spending their days being bombed from the air and shelled lightly from land. At night they are under more bombing and very heavy shelling from battleships and cruisers. So much heavy stuff is falling that the eight-inch guns of the cruisers seem to sing soprano to our fellows.

Among the men who have dug in here are Lieutenant W. O. Adams of Coronado, California, paymaster, and Sergeant Miles A. Williams, paymaster clerk, who wandered into the battle carrying a canvas beach bag full of several thousand crisp dollar bills. The men rode in a transport plane, sitting on the tops of barrels of gasoline. Incidentally, these planes, familiar externally to civilian air passengers in the United States, have been threading in and out of the war for the Solomons laden with supplies. They ride with a prayer and parachutes.

Bombs began to fall as the paymaster and clerk stepped out at the airport. They dropped their little brown bag of dough and dove head first into the nearest foxhole. As soon as the bombing stopped, they carefully retrieved the money unimpaired and found to their distress that nobody wanted it.

'Keep it for us until we get a post exchange up here where we can spend it,' they were told. Not even bombs seem to find money useful, the bag being unhit.

Our men spent the day and night sitting on the edges of foxholes, waiting to dive when such a proceeding was useful. Mess went on more or less as usual, the men being called from their foxholes in groups of ten and diving with the mess crew when the stuff began to fall. The stew was dusty, but the dust only made it taste more salty than usual.

With the Navy out somewhere along the burning edges of this battle, obviously effectively, and Japanese shock troops being turned, it is hoped, into shocked troops, more confidence was felt today than at any time this week. However, the Japs are using picked, commando types of soldiers in this battle and the amount of shock necessary before they get shocked involves practically electrocution. It is not a tingling little massage job that our Marines and infantry face now.

CHAPTER 5

Round 3

★

CHAPTER 5

GUADALCANAL, SOLOMON ISLANDS

The battle for the southern Solomons underwent a complete reversal in character in the course of October 15 and 16. Our Navy was up to something obscure, but apparently deadly, in the 'wild world of waters' surrounding the islands, and the Jap fleet which had been covering troop landings with a heavy barrage from air, sea, and land suddenly vanished.

Our troops lifted their haggard faces from fox-holes Friday to find themselves surrounded by silence which seemed to ache as with a hangover from the explosives thundering nights and days before. Then the Jap troops huddled in the jungle began to taste the same kind of hell as our troops have been burying their faces in. Navy and Marine Corps dive-bombers and fighter planes plowed them up, harrowed them over, and blasted flat whatever the turned earth revealed. Army Flying Fortresses came over in low-level sweeps, dropping

high explosives on the stores and supplies which the
Japs had thrown hastily from ships onto the beach.
Our ships added to the hell by shelling them heavily
from the sea.

This was exactly what the Japs were giving us
during the previous days of this battered week, and
the mutter, rumble, and earth shocks, now twelve
miles distant, came to our troops like the noise
made by some friendly giant, socking and booming
over and over again.

Jap supplies are still sprawled in untidy heaps
along the beach at the edge of the jungle this morning. The Japs have now had six days to haul them
under the shelter of trees, proving either the effec-
tiveness of our barrage or, less likely, the scarcity
of their manpower.

It is still too early to predict the course of this
third battle of the Solomons with any certainty.
The Japs have been accumulating land forces
steadily and furtively for three months, sneaking
them in under the cover of darkness. These forces
have not yet begun to drive against our trenches,
protecting the Henderson Airport fighter strip. A
few weeks ago, Jap troops filtered, poked, and tip-
toed their way into the Marine-held area which is
now known locally as 'Bloody Knob.' Approximately two thousand Jap bodies have been removed

from there subsequently for burial to avoid pestilence.

However, although nothing is certain here, not even death and taxes, one thing seems rather clear — the Japs who started the war with a slam-bang rush have been slammed back into caution. They are no angels but they seem to fear to tread. From the store of slender information that can be acquired by a reporter inquiring among bombed foxholes and gunned cockpits, it would seem that the Japs went into battle with two task forces. A task force usually consists of a portable airport, otherwise known as a carrier, surrounded by a vast swarm of protecting vessels, ranging from battleships to destroyers.

Since this war is being fought to gain what in peacetime are known as islands and now are called 'unsinkable aircraft carriers,' transports filled with troops to hold the unsinkable carriers once they are 'won,' and tanks, artillery, planes, and supplies to help them, usually nest amid the task force.

Well, the Japs seemed to have had two out in the water beyond here, both together comprising a force as large as the one that figured in the Midway battle, but, unlike Midway, the Japs kept their main striking weapon, their floating carriers, out of range of our land-based forces, reducing their efficiency but saving lives. Attacks are not made good by lifesaving tactics. This is a cruel truth of war, which the Japs seem to have learned as well as the Germans, so it does not seem too rash to assume that the Japs' losses, at least in carriers, have been so great they cannot afford more, and that this much of a turning point has been reached in the Pacific War.

If the nature of the war here has been puzzling you, then you have plenty of company, but it seems primarily to be a naval free-for-all in which every weapon known to man, from the most primitive to the most modern, plays a part. This is probably the most extraordinary naval war in history. Admirals not only have to use carriers, battleships, and submarines, but also field artillery, machine guns, rifles with telescopic sights, machetes, and even bows and arrows. Our troops do not carry bows but the natives do, and their jungle tactics will form A.E.F. legends of the immediate future just as the tactics of the giant Senegalese in France formed A.E.F. legends of the past.

Incidentally, one of the most heartening aspects of the war in this remote corner of the world is that wherever American boys go, they win over the natives to their side, not by any planning or propaganda, but simply by being Americans, meaning democratic, willing to sit down with the natives in their huts, admire their prowess, learn their arts, and share their knowledge. The commanding officers at Guadalcanal have learned to restrain civilized emotions when barefooted natives come trotting into camp. Exactly what the natives think of a civilization their first contact with which is the mechanized war bursting around them, remains locked impenetrably, because of language difficulties, beneath their frizzing and often glowy red-thatched skulls. But that they think our way about who the enemy is seems plain.

The reason for the profusion and confusion of weapons in this naval war is the fact mentioned previously — that this sector of the war is being fought primarily for islands to be used as unsinkable aircraft carriers. Not only must the jungle-covered land be held and made useful by the Navy's construction battalions and the Army's engineers, who labor like ancient Romans — under such long-range fire as the Romans never knew — to stamp permanently the imprint of civilization upon a primeval wilderness, but the waters around the islands must be held and the lines of communications must be batted through and preserved over areas of thousands of square miles. Water and over-water communications is Navy business, and they attend to it with a hard-hitting vigor which is earning the respect of all the navies of the world, including the Japanese.

This, at least, is the view of the war obtained by a man lifting his head from a foxhole every now and then to take hasty peeks and squints through the fog of flame-licked smoke. A bird's-eye view of the battle area of Guadalcanal, obtained while perched in the nose of a Flying Fortress on a bombing sweep of the Jap beach-head, made at better than three hundred miles an hour, reveals evidences of Jap submarines, some say as many as forty, prowling along the lines of communications of the islands. The airport and fighter strip, looking like white wounds laid over the darkly flowering earth, show that all runway damage caused by Jap attacks has been repaired.

The night shellings by Jap warships were of intense ferocity, and bombings were featured by Jap formations of as high as eighty-five heavily laden planes. But most of the damage seems to have been done to the cleared areas around the runways, these areas being pocked, pitted, and pimpled like the relief map of a volcanic area. Those men who had duties which did not compel them to be around the airport runways slept in the front-line trenches a few miles away; as so many millions have discovered tragically in this war, the front lines are the safest

place to be. But very many other men in Guadalcanal lived right in the midst of volcanically thrown,
chewed, and flung-up earth, and yesterday afternoon
they could be seen standing all over the runways,
having crawled from the hit-at and gashed burrows.
These invincibles, who the Japs must by now believe
are not made of flesh but of some shatter-proof composition, looked very small from above, hardly more
than black dots. Their tiny silhouettes were outlined as sharply as flags against the white runways,
and they looked like flags of some desperate cause
as they stood whippily with their feet apart, hands
on hips, with an assured kind of swagger looking up
silently at American bomber squadrons roaring
overhead.

The island seemed as quiet as held breath. There
was no anti-aircraft fire, proving either that the Japs
have not yet had an opportunity to set up their batteries until their forces were ready to strike, or they
lacked guns or ammunition. Machine-gun sniping
fire cannot be detected from the air. However, it is
likely that the battle which started as an old-
fashioned 'better 'ole' kind of war has now got even
more old-fashioned and become a ruthless, tracking,
potshotting, Indian kind of war. Our fellows are not
the sort to let the snake-filled grass grow under their
feet.

By now (October 19) it is pretty well believed among the high command here that the Solomons situation is once again 'in hand' and that the back of the high-powered Japanese effort to throw us out of here has been broken.

Jap troops have retreated sixteen miles down the beach under pressure from our forces. They just picked up and went without making any kind of stand, influenced, it is assumed, by the fact that combined Navy, Army, and Marine bombing, shelling, and strafing operations from land, sea, and air left them little material with which to make a stand.

The Japs have only one artillery battery left within hitting distance of our air field. A cat-and-mouse game has been played with all batteries they have set up and this is the last mouse left. He is a tricky one. He never fires until the dive-bombers and pursuit planes hunting him have their tails turned. Then he lets a peep out of him and quickly ducks back into his hole.

It seems to be a five-inch battery. The Marines at the airport treat its shells the way city-wise pedestrians treat taxicabs — with caution, but without nervousness. One hears a dull boom, followed quickly by an echo which sounds like a thumping little brother. 'Down!' calls somebody, and everybody falls flat on his face, not with any particular hurry, because some seven seconds elapse before the shell can be heard whistling on its way to work. The last thing that can be seen as the men hit the dirt is exasperated planes overhead making tight turns to try to spot the gun. They fling around angrily as if stung in the tail.

The Jap gun is firing at extreme range and falling about two hundred yards short of what obviously is his target. He has left a trail of shell holes in a straight line, looking like giant footsteps of some oriental version of Polyphemus. Orientals do most things backward from us, so their Polyphemus could have one foot instead of one eye.

Anyway, the shell holes come firmly, implacably, each a little more forward from the other, until they begin to wobble. This, I was told, indicates to artillery experts that the boundary of the gun's power has been reached. But to the unpracticed eye, the flurry of shell imprints looks as if this one-legged giant had been stamping with girlish petulance at being balked from advancing farther.

If the gun intends to do any good, it will have to come from its hole and move up. When it does this, our fellows are sure to destroy it.

So the situation here has returned to what is normal in Guadalcanal. Our men are bush-fighting the Japs. 'Lousy Louie' comes over nights and drops an occasional bomb. A Jap warship sneaks up, throws rocks, and skedaddles past. This means that we retain the upper hand in Guadalcanal, but the Japs are still clinging to our hand tenaciously with their finger-tips and are doing their best to prevent us from hitting them a real knockout blow with that hand.

By October 31, it will be three months since the first Flying Fortress dropped the first American bomb on Guadalcanal.

We took it August 8 and have held on to it ever since. But the Japs, while unsuccessful in two all-out *blitz* drives to take the island back, have been successful in preventing us from putting it to proper use. This is the blunt truth. The southern Solomons is a base. But the Japs have insisted on making it a battlefield and have sent enough matériel down to make sure that that is what it remains.

We, on the other hand, have consistently asked the men here to fight against overwhelming odds. No doubt, a report of exactly what the odds are would be of benefit to the enemy, but the enemy knows by this time that six Flying Fortresses were sent October 12 without fighter protection to bomb thirty-six Japanese warships in Tonolei Harbor, off the Solomons. They had to plow through swarms of Zeros to get to their target and plow through swarms more to get away.

This is not the whole story. The day after our effort against their ships, the enemy made two separate air raids — not on anything so punishing as thirty-six warships, but on the airport at Guadalcanal. The first raid was made by twenty-two Jap bombers, the second by fifteen bombers with strong fighter support. This is not shooting off firecrackers in the wind. This is the kind of concentrated power Germany was bringing to bear on England when she was making her bid to crack that base of future operations.

Incidentally, the evening our six bombers returned from their raid on the thirty-six Jap warships with a score of hits on two ships, I spent some time with the unwounded members of one of the crews listening to music and news broadcast from San Francisco. The music unbuttoned their nerves, but when the cheerful baritone voice began peddling the news in tidy little, neatly packaged paragraphs, the nerves buttoned right up again. He announced heartily that a hundred and fifteen Flying Fortresses accompanied by Spitfires had bombed something or other in France — a railhead, as I remember it; anyway not thirty-six warships. The men listening looked at each other silently and asked questions of each other with their faces.

Finally one said, 'Holy crock!' and another said: 'Fighter support! Think of getting fighter support! Bombing with fighter support would be like coasting down a hill in January.' The fighter planes here, because of a number of necessities, have had to limit their support to dive-bombers and torpedo planes.

Despite these odds, about four hundred Jap planes have been destroyed by us, on the ground, in the air, or by attacking carriers. The air command here says that operational losses always at least equal combat losses. In the same period, the Japs have shot down two of our Flying Fortresses. We have lost others in collisions with Zeros and on the ground and have lost considerably more fighters and dive-bombers, but nothing comparable to the Japs.

In the same period, naval engagements have been frequent. Last week was not typical, but there have been others like it, and in the course of this one week the Japs lost three heavily laden transports, two cruisers, and four destroyers and suffered direct hits on one battleship and a cruiser. In achieving this, our ship losses were two destroyers.

The Japs are proving they cannot take such losses in their stride. At the beginning of the war last December, there was an interval of one year between the time a plane left a Japanese factory and the time we shot it down. In ten months since December 7, the interval has been shortened to six weeks. A plane shot down yesterday over Guadalcanal still had the paper wrappings glued to its propellers. They had not had time to wash them off before throwing the plane into combat.

And today, as every day during this week, bombers returning from search missions report Japanese striving desperately to salvage motors and parts from flying boats that have been shot down and have failed to sink.

These losses are our gains in the battle of the Solomons. But while we have gained much, we have not won a victory there. Nor shall we until we bring in enough matériel to roll the Japanese out of range of Guadalcanal and give us a chance to consolidate this torn and pulped husk of land as a base. They will not go back easily, nor will they retreat voluntarily to spare themselves further losses. They know that as they retreat, we advance, and the bleeding process begun in the Solomons will find no tourniquet in Rabaul.

CHAPTER 6

'Git or Git Got'

★

CHAPTER 6

★

'Git or Git Got'

FROM A BASE IN THE GUADALCANAL SECTOR

Throughout the day and night in this enormous arena, life moves along in familiar bursts and spurts, trying nervously to keep a step ahead of Death. Every now and then during the day, Death pulls abreast of us, and it is touch and go there for a minute or two while death fumbles among the living and breathes on us like an animal and paws us over and huffs and stamps around, trying to make up its mind whom to take. A few times during the day, on this island or that, in this wave or another one, Death makes up its mind and takes some of us, but there are other times when it can't seem to make up its mind at all and just passes everybody over.

The fellows kiss off this kind of day by saying it makes 'a lot of good Christians' out of those who survive and let it go at that, while a reporter just fiddles around, itching for something new to write home about.

I have been sharing a tent for the last three weeks

73

with Lieutenants Loberg, Thurston, Spitzer, and Mitchell. We came out to the war together, from Honolulu. These fellows and their crew are all veterans of the war here, having come out of the battle of Midway to drop the first bombs on Guadalcanal July 31. They've had a long, merry life. They call me 'Scoop,' because that's what they promised me, a scoop, if I stuck with them.

At 2:35 this morning (October 23) I woke up to see a Japanese star shell hanging in the sky over my head and rolled right through the mosquito bar and off the cot, landing with my face buried in my arms. I did all this in a single motion which has come to be known among us as the Guadalcanal twitch.

I landed very close to Lieutenant Loberg's cot. 'Scoop!' he called. 'Hey, Scoop!' I lifted my head. I could see him very plainly. He had raised himself on one elbow and was staring down at me with a sleepy, worried look. 'Wake up, Scoop!' he said. 'They're shelling!' I asked him what the hell else he thought would be making me sleep on the ground, if not shelling.

Then Operations decided that Lieutenant Loberg ought to take his Flying Fortress up and get whatever was shelling us. He was scheduled for a restful search mission that day, having barely returned from a really frantic bombing run over a covey of

74

Jap warships that had been busting in where they didn't belong. And I had decided to go along to complete a story I have been working on, a description of the most unusual thirty-seven hundred miles of cohesive battle front that have been established between the Japs and us. I had already flown over a substantial portion of the front line and the extra four or five hundred miles covered in a search would add to the picture of it.

Lieutenant Loberg thought he'd circle around awhile and look for whatever it was doing the shelling and not find it, just scare it away and then be ordered by radio to swing off into his search. So I went along with him on the first mission in order not to be left behind on the second.

The Japs had been firing salvos of four-inch and six-inch shells. They had been hitting coconut trees and so forth and the fragments had been showering down along with coconuts and tree limbs. We went blundering and grumbling along through the blacked-out darkness and the trees, Lieutenant Thurston being particularly sore because the Japs had woken him up from dreaming about a girl he knows back in Indianapolis. We passed a fellow wandering around with a dazed expression holding a chunk of shrapnel as big as a bucket that had landed near him and nestled up alongside of him, and passed lots of fellows trying to finish out their sleep in foxholes. We also went by an infantry sergeant standing over his cot, putting his head against a big shrapnel hole where his head had lain a minute before, and by the runway there was a colored soldier standing guard and saying in a high, girlish voice, 'Lawdy me, Ah tho't Ah so' mah number on that shell, shore.'

We didn't find anything. The Japs ran away. We circled a long time over the water, hunting by the light of a big tropical, glistening moon and then by the light of dawn, and finally were ordered back in — to finish our sleep, we hoped.

The big Fort sat down on the runway as softly as a falling petal — greasing it in, they call that — and Lieutenant Loberg put on the brakes.

Then we found out there were no brakes, and there we were, all the bomb-loaded tons of us, tearing along down that runway with nothing to stop us except what Fate might happen to think up. Loberg had an idea that if he could get enough brake on one side, he could ground-loop at the end of the runway and whip the plane around and run back over the runway, and keep doing that, go back and forth that way, until we had lost all momentum. But there not only wasn't enough brake for that, there wasn't any brake at all, and what finally stopped us was the wing of another plane, a parked one.

It came right up against our wing and both wings wrapped around each other like loving arms. But we weren't thinking of that at the time. There was a hundred-pound bomb on the other wing and our wing had snaked it out, pulled off its arming wire and moved its firing pin and sent it bumping along under us. When we saw that, we didn't think of anything at all, just held our ears and waited for the explosion.

The ordnance man from whom I tried to find out why the bomb didn't go off just said we were damn fools to hold our ears. 'If that bomb had gone off,' he said, 'you never would have heard anything.' This didn't explain anything, but it seemed to satisfy him.

Intelligence still wanted that thing found that had been shelling us. All the other fellows were busy doing something else, so we got into another plane and went hunting over the sea. We hunted for about five hours and at 12:28 of a tropical noon packed swollen with weather we found something — not what we were looking for, but an acceptable substitute.

At the time, there were patches of steaming sun lying breathless on glassy water and patches of squalls and cloudbursts rising in huge thick pillars over heaving seas. We had been searching low and had found nothing and had climbed up to about six thousand feet to give whatever Jap wanted to a chance to sneak in under us.

Lieutenant Loberg called for battle stations. Everybody started running to where he belonged and I looked out the window. Far below us there was a PBY, one of our own Navy planes. Its big broad wings sticking straight out made it look like some bird coasting on wind, and yet the way it turned and lifted and fell gave it an appearance of floundering and an appearance of anxiety. Near it was the Jap flying boat — the best four-motored bomber the Jap navy has — and it, too, looked like a bird, but a bird bunching itself for the kill, its four propellers glinting and curving and cutting like claws.

The PBY pilot came back later and said he was chasing the Jap at the time, but a PBY chasing a Kawanishi 97 is like a chicken chasing a hawk, and the way it looked to me it was the Jap who was doing the chasing. Anyway we muscled in and took the fight over, never seeing the PBY again after that first glimpse of it.

We dove straight down so rapidly that my knees buckled under me. I fell on my knees as if hit and tried for a long time to lift myself up, but I couldn't against the tremendous weight of gravity. Nobody knows how many times the weight of gravity was multiplied by that dive, everybody being too busy to notice. But I could feel my cheeks pull far down below my eyes and could feel my eyes sticking out of my head and my cheeks resting on my shoulders, and everything that was inside of me pressed into a bundle and forced into my lap.

The bottom turret was working. The nose bucked with its chugging, like the handle on an electric riveting machine. I noticed it while still struggling to get off my knees, and the last thing I saw before the whole air turned dirty gray and the blackout began was the pointed glass nose of the plane shining in the sun, bucking and tossing.

When we got on an even keel, I could stand up again and look out the window. We were in a squall. The window was black with wind and clouds and rain, the rain drummed all over us with a sound like running feet. The plane was being thrown around so much that I had to hold on to the navigator's table with two hands to keep from falling.

'What happened?' I shouted to Lieutenant Spitzer.

'We lost him,' he shouted back. 'He ran into a cloud to hide.'

And we, it seemed, had dived in head first right

after him. At every window men stood straining to see through the scudding, flocking mass of wind-blown water trying to find the Jap. Then we hurtled clear through the cloud and into blinding sunlight and there the Jap was, right alongside of us, maybe fifty feet away, the red ball that means ice skating to us and the rising sun to them plain on his tail.

Every trigger on every gun in both planes was pressed at once and held. Thousands of bullets criss-crossed through the narrow spread of air. The planes rocked along side by side. It was a fantastic spectacle. Our plane shuddered under the impact of bullet after bullet and teetered and buckled under the blasts of its own guns.

The chuggings of the guns became a single monstrous chugging. The thousands of explosions seemed one vast unending blast. I could see the Japs clearly, four or five or seven, something like that, small, shrunken-seeming figures huddled up over their guns. I could see a cannon firing at us, smoke blowing from its open mouth like frosted breath, and I could see our red tracer bullets pelt like darts into the Jap and ricochet off his armor and streak wailing straight up into the air.

I could see our men and their men thrust their almost naked selves right into that terrible fire.

They were bent over as if before a high wind and all the faces were wrinkled and gray as cloth, but they stood to it and kept at it. Air fighting is not like land fighting or sea fighting, where a man can take cover sometimes. There are no trenches or foxholes or gun shields on a plane, just a thin skin of metal and glass, mostly glass.

The Jap made a tight turn and we followed suit. It was a delicate maneuver, the Jap having a mortal sting in his tail. To keep away from his tail and give our nose and side guns a chance to work him over, we had to turn inside him. This many believed impossible to do without breaking a B-17 in half or putting it into a spin, but Lieutenant Loberg did it and the Fort did not break in half, but just kept on raking.

Then the sun was blotted out and the whole wild scene was blotted out by an even wilder one as a cloudburst fell upon us like a mountain of loose rocks. Lieutenant Spitzer had put on his sweatshirt when we had gone up into the upper air before sighting the Jap. Now he stepped away from his gun. He looked as if he had just stepped dripping from an ocean of perspiration. 'Oh, me,' he said. 'Oh, my,' and pulled and pinched at his sweatshirt and blew a current of air from pursed lips up past his nose. I looked at him startled. It seemed remarkable that he should utter such mousey words at such a time and stand plucking and pinching at himself like some woman, some plump churchgoing woman being fretful about a hot afternoon. After a moment, Lieutenant Spitzer took off his sweatshirt and threw it on the floor and went back to his gun.

The Jap had dived into the cloudburst, either to lose us or to maneuver himself into a position to kill us. We lost him five times in the fight, sometimes for three and four minutes at a time. The Jap was very smart and also very brave. He seemed to be a veteran of this kind of fighting. He knew just where to come in on us and just when to hide and how to hide. But Lieutenant Loberg and Lieutenant Thurston are veterans, too, and brave, too, and seemed to be smarter than the Jap, because they outthought him every time. They guessed every time where he was hiding.

They had the help of the crew, 'marvelous help from them,' Lieutenant Loberg says, who kept looking to the very last flicker of the disappearing Jap and reporting what they saw of his maneuverings. And every time the Jap plunged into a cloud and went racing along behind the cloud as if it were a board fence, we went racing after him and caught him just as he was breaking into the clear and hit him.

The Jap kept very close to the water to make sure we wouldn't come in under him. He has no guns there and we could tear his belly open with our top turret. So the fight was so low in the air there would be no escape. A hit on the controls for either of us meant the end. There would not be time to bring the plane out of a dive or spin. There would not be time for anybody to bail out or even get through the escape hatch if we hit the water. This was kill or be killed all the way through, one of those arrangements known locally as 'git' or 'git got.'

I kept thinking of that and kept wishing that our crew would not be so damned smart every time the Jap got lost. I would say good riddance to him, and let him stay lost. I was willing to say good riddance the moment we saw him. I was working on a different kind of story altogether, an interesting and constructive story, not this messy, firelit mechanized adventure stuff at all. But I had nothing to say in the matter, and even if I had had, I couldn't have made myself heard above the roar of the four motors and the roar of the storms and the roar of the bullets.

No, it isn't until now when I am back safe on *terra firma*, and repeating the old saw about the more *firma* the less *terra*, that I begin to appreciate

83

the great amount of thought involved in the fight. It was not only a question of fighting in the kind of weather that no pilot in his right senses cares to meet and of stunting and half-rolling and power-diving and turning tight in that weather, straining the plane past the charted limits of endurance, but it was a question, too, of trying to remember what the Army or Navy Intelligence had said about the guns on this spectacular Jap plane and trying to remember which of your own guns had been shot out. Also, it was a question of maneuvering, in the midst of all that turmoil with Sergeant Paul Butterbaugh pumping cones of bullets right across Lieutenant Loberg's and Lieutenant Thurston's eyes; of maneuvering, too, to bring your strong points against his weak points and keeping your weak points away from his strong points.

I don't remember exactly when Lieutenant Spitzer got burned. We were taking bullets in the nose fairly steadily through the fight. They were flicking, whisking, and smacking all around us. I remember once he rounded his lips as if howling, but I couldn't hear the howl in all that noise, and he went right back to his guns, so I didn't think any more about it.

But Lieutenant Mitchell was hit by fragments from an armor-piercing bullet that buried itself deep in the side of his machine gun and made a smack that could be heard above everything else. We both looked at him, frightened. He was standing dazed over the gun. His head was bowed over it and his face hung down from its bones, all stunned and loose-looking. I tried to get to him, but the plane was bucking terribly, and I couldn't move across the few inches of floor without falling. Lieutenant Mitchell stood trying to balance himself, his head stooped and rolling slackly on his neck. Then he tried to work the trigger. No bullets came out, and he tried to lift off the top cover and see what was jamming it, but the top cover had jammed, too, and wouldn't come up.

I thought he must be all right if he was fussing with the gun. There were plenty of other guns for him to work in the plane. Then some time later, I don't know how long later, I noticed he was standing alongside me. He put his lips close to my ear so that I could hear. 'Where am I hit, please?' he asked. His voice sounded very soft.

Blood was coming down his eye and dropping down his neck into the golden hairs of his bare chest. I wiped the blood away with my finger to see how deep the wounds were and saw they were only gashes. I told him that. 'My foot hurts, too,' he said. 'I can't stand on it.' He seemed to think it was bad luck that the only gun in the nose he could work lying down had been shot out.

Twice we passed right over the Jap, so close I could see the jagged bullet-holes we had torn in him, and I lifted my feet gingerly, one foot at a time, hardly knowing where to step as I looked down at the floor, expecting a spew of bullets and cannon shells to come up through it. Then, at 1.01, exultantly Lieutenant Spitzer shouted, 'He's smoking! One of his motors is gone!' I could see the propeller windmilling idly.

Then Lieutenant Spitzer, who was still working his two guns, shouted, 'He's down.' Lieutenant Mitchell was sitting quietly in a corner on a parachute pack. I turned and asked him if he'd like to have a look at the Japs who had hit him. There was an aching silence in the nose of the plane. Anyway, it seemed like silence. We had all been deafened by the noises of thousands of explosions.

Lieutenant Mitchell tried silently to get to his feet. I helped him up and held him as he stood looking out the window at the somber spectacle on the sea below. I had a camera in one hand and held up Lieutenant Mitchell with the other. He was steadying himself against me with one hand and trying to clear the blood out of his eye with the other. He wanted to see plainly. Lieutenant Spitzer had dropped his guns for a movie camera. He was grinding away on that with one hand to provide headquarters with proof of the kill and was lending his wounded friend the support of his spare shoulder.

The three of us stood like that in the smother of quiet, pressed against each other, looking out the same window. The Jap plane was burning like a tanker. There was the same swollen oval of flame laid like a blister over a sea as smooth as skin. The flames were an orange-colored red and rolled up in big, unfurling billows, flapping like a flag into clouds of mole-black smoke.

There must have been twenty acres of flame when we first passed over. In the center of the flames lay the frame of the Jap plane. It was buckling before our eyes. There was nothing left of the plane but the frame, and the frame looked skinny and black like the bones of a skeleton disintegrating. Two small, black objects that could have been men or maybe were just round bits of rubble or débris had been thrown clear of the plane and were on the edge of the oval of flames. They were either trying to get away from the flames or maybe just being swirled away by the currents created by the heat.

The flames were spreading rapidly and the smoke

was mounting. We passed the first time at about two hundred feet, according to the altimeter, and the smoke was over our heads. I asked Lieutenant Mitchell if he would like to lie down now. He didn't answer. He kept looking out of the window and trying to clear the blood from the one eye. We circled and came back at five hundred feet. The smoke had mounted even more rapidly than we had. It was as thick as fungus and mushroomed out high above us. The whole thing, the blast furnace on the bottom and the smoke and everything, looked like Pittsburgh breathing on the planes that go by there. The skeleton of the Jap plane had disappeared completely, and the oval of flames had covered the place where the two black little objects had been.

For one minute short of a very long three-quarters of an hour, while battling through about a hundred and fifty miles of big, black, cracking cloudbursts and squalls with air currents in them, or, as they are called, thermals, each of them calculated to hack any ordinary plane apart, the Flying Fortress had used the pursuit and interception tactics that are standard for fighter planes in a dog-fight.

The pilot, First Lieutenant Ed Loberg, twenty-seven, is a former farm boy out of Tigerton, Wisconsin. The co-pilot was Bernays K. Thurston, an accountancy-minded, guitar-playing, blues-loving youth of twenty-three from Indianapolis. Together, with a 'Now, you take it, Beekay,' and a 'Now, you take it, Ed,' as they tossed the controls from one to the other when the enemy turned from one side to the other, they threw the Flying Fortress through whole storming cauldrons of air. They seemed to be under the impression that their huge plane was one of those Beebee-size, bat-and-ball P-shooters they have around here. They made it do all the things that only a pursuit plane is supposed to have been built to do. And they did it up and down and along and through the kind of weather that a Spitfire or Focke-Wulfe or Zero or one of our 'P' series of planes wouldn't care to fly into — even if it had nothing more on its mind than to stay up.

During the duel, the Fort that I was on, with a bullet in one of its motors and two holes as big and flowering as Derby hats in its wings, made tight turns with half-rolls and banks past the vertical. That is, it frequently stood against the sea on one wing like a ballet dancer balancing on one point, and occasionally it went over even farther than that and started lifting its belly toward the sky in desperate efforts to keep the Jap from turning inside it.

It made numerous spiral dives and, at least twice,

although in the excitement of the battle it was difficult to keep accurate count, it dove vertically like a dive-bomber and pulled out so rapidly that a curtain of soot-colored gray, indicating the beginning of a black-out, dropped over my eyes.

Throughout the entire forty-four minutes, the plane, one of the oldest being used in this war, ran at top speed, shaking and rippling all over like a skirt in a gale, so many inches of mercury being blown into its motors by the superchargers that the pilot and co-pilot, in addition to all their other worries, had to keep an eye on the cowlings to watch for cylinder heads popping up through them.

The Jap, with a somewhat lighter plane, counted on greater maneuverability to help him. But he must have counted wrong because he is now dead.

Throughout the battle, I stood in the glass-enclosed nose of the Fort with the navigator, First Lieutenant Robert D. Spitzer, twenty-six, of Anderson, Indiana, and the bombardier, Second Lieutenant Robert A. Mitchell, twenty-four, of Washington, D.C. With five guns shot out in other parts of the plane, the nose had to do more than its share of the fighting.

All other gun stations had narrow escapes. Corporal Everett Gustafson, twenty-four, of Malden, Illinois, who was at the tail guns, had a Jap 7.7

explode one of his own bullets exactly in back of his lap. Sergeant Ellsworth Jung, twenty-two, at the waist guns, and Staff Sergeant George Holbert, twenty-five, of Lamar, Colorado, who had run up to help Jung when his own bottom turret had been shot out, had Jap bullets passing between them and between their separate legs. Staff Sergeant Edwin Smith, twenty-three, of Petersburg, Indiana, and Technical Sergeant Paul Butterbaugh, twenty-five, of Altoona, Pennsylvania, had narrow brushes, too.

All were sticking themselves head-first into the line of fire without anything between them and the enemy's guns except air and a composition glass, that being the way airplanes are constructed. All were pumping all the lead they could, but the nose took all the hitting the Jap dished out and did the bulk of the hitting back.

At the time the Jap hit the water two hundred feet below us, both Lieutenant Mitchell and Lieutenant Spitzer had been wounded and were in pain, and something had happened to me, too. An armor-piercing bullet had hit the machine gun Lieutenant Mitchell had been working. The explosion had stunned him and fragments from it had gashed one eyelid, the side of his neck, and his hand and one foot — nowhere seriously, we discovered later.

Lieutenant Spitzer had been burned on his legs in five places by five hot shells without any of them breaking the skin. A bullet had plowed through a metal ready-box loaded with ammunition over my head, starting a small fire under a high-pressure oxygen cylinder on which I was resting one knee.

However, fragments from the explosion, in the midst of which my head seemed at the time to be nesting, had passed me by — apparently, to be literal about it, too close for comfort. I say apparently, because I can't be sure about anything but the lack of comfort. There was a smarting sensation along a line under one eyebrow and a wider one across my chest, but I have only two dot-like scars by which to remember it.

And that was the last news in the encounter. It is news in this long-range war of machines where steel is thrown in the darkness from twenty miles away and bombs are dropped on you by planes so high that only the vapor of their passing can be seen — anyway, it's quite rare nowadays for a man to be able not only to kill exactly those who have hurt him, but also to stand up and watch them die.

On the way back, after we had stuffed and powdered Lieutenant Mitchell with sulfanilamide and bandaged him and made him as comfortable as we could and had jellied over Lieutenant Spitzer's burns, I wanted music. Music seemed to be the only thing that would do, and Lieutenant Thurston obliged over the interphone system, singing a marvelous little song he had picked up somewhere:

> We're marching, we're marching,
> Our brave little band.
> On the right side of temperance
> We'll all take our stand.
> We don't use tobacco
> Because we do think
> That all those who use it
> Are liable to drink.
> Down with King Alcohol,
> A-A-A-Amen!

It was not the song of warriors returning drunk with victory, but everybody joined in the second chorus, particularly in the 'Amen.'

CHAPTER 7

Battle in Three

Dimensions

★

CHAPTER 7

Battle in Three Dimensions

FROM A BASE IN THE GUADALCANAL SECTOR

The fourth battle of the Solomons, which gave Admiral Halsey his baptism of fire as Commander of the South Pacific Force, was actually a development of the third and followed hard upon its heels. The week of lull in between was an uneasy lull, the Jap remaining in touch with us. At least he kept trying to touch us with his high-explosives encased in bombs and shells and succeeded as often as he failed.

This fourth battle was typical of all the battles the Jap has fought to regain the Solomons and, for that reason among others, will be described here in detail.

The battle was fought over an area of two hundred and fifty thousand square miles, the bulk of it covered with water, and lasted five days and five nights. These days were as fierce, bloody, and uncertain as all days of pitched battle are.

As far as this reporter can determine with his limited knowledge and the limited view he was able to get from a fast-moving seat in the arena itself, this battle was a work of considerable art on our part in strategy and tactics. Certainly it had in it all the chaos and tragedy and exultant soarings above tragedy of all works of great art.

Our side wrote the story. The Jap supplied terrible chapters in it, but we were in charge almost all the way through and in charge of the ending. The ending of the battle was that the Jap failed to achieve his objective.

In addition to all their other difficulties, both sides raced the weather. Nature entered the arena with a front of its own, a black and storming pile of weather which is known even technically as a front. This front, like something created by Thomas Hardy, began advancing upon the battle area as the Jap fleet steamed out for action. Meteorologists, or, as they are called here, weather officers, charted it anxiously. It had a long way to go, but it sent speeding out ahead of it scouts, points, screens, and spearheads in the shape of squalls and cloudbursts, Nature's main force following on behind.

The men, standing up tall to this eruption of Nature, used it to conceal their maneuverings at first. But on the fifth day in the early hours of the present morning, Tuesday, October 27 (Monday, October 26 in New York), Nature's front took charge of the arena and set it boiling with a fury of its own. Admiral Halsey is known to his men here as a 'rough brush.' But not even a rough brush can sweep such a storm as this one out of men's eyes, and in the blackened day that is now following the black night, nobody, not even the tallest man, can see anything it would be useful to kill.

At the moment this is written, the various American staffs engaged in the battle are trying to add up thousands of eyewitness reports and determine how roughly the Jap was handled in his fourth effort to throw us out of Guadalcanal. When they have arrived at the score, they will radio the news home and a communiqué will be given out, which you will have read long before you can read this.

For reasons which are not only irritating, but seem unnecessary, the Navy denies the press here any access to rapid communications and so handicaps it seriously in its war job of helping to wake up our people to the fact that we are not going to start to fight when we get ready, but are fighting now, bloodily and desperately.

I don't know whether this dispatch will reach you in a week or two or three months or even whether it ever will reach you, but I do know, having been on the mainland and out of the war as recently as last September, that people cannot be stirred by facts as old as these will be by the time they reach print, and cannot get any sense of continuing pressure on their emotions unless there is daily communication between them and the battlefield.

I had no idea until I got here that a war was going on in this sector — the only serious-minded war the Japs are fighting anywhere in the world at present — and that shots were being exchanged daily and men were being killed and wounded daily. And that, for instance, the handful of Flying Fortresses engaged in this area have had exactly one day since July 31 when no enemy contact was made and no job of bloodying him up or getting bloodied up was done, and that our Navy has been in action during the same period with the same, if not greater, intensity. I had no idea of this because the press, where I am accustomed to getting my impressions of what's going on in the world, was not allowed any means of telling the story.

And, to tell the truth about it, this dispatch is being written by a reporter who has a story to tell and feels that telling the story would be useful to his country's effort to win the war and can't think of anything to do under the existing circumstances but tell it, even though he will have to trust to luck to get it to where it can be heard at a time when people will still be interested in hearing it.

This morning, in the midst of a widespread storm which will last several days, the Jap fleet broke off the engagement with ours and withdrew to its bristling fortresses in the North where we are not yet prepared to follow. The Japanese land forces on Guadalcanal, who had been trying to drive the Marines and Army infantry out of their way with everything they had, including fourteen-ton tanks and bayonets, and had succeeded in making something like a break-through, have now retired to positions where there is no point in our paying the cost of following them.

There is no point in following them because they are now relatively harmless without reinforcements and will die peacefully of starvation, as a cohesive army at any rate, if their sea lines of communication are kept broken. The Jap land forces lost nine tanks in the battle and approximately twenty-one times as many men as we did and have retreated to territory that we have no use for.

It must be remembered that the fight for Guadalcanal is not for the hundred-mile island, but for the air facilities on the one tip of it and the sea roads leading there. Those air facilities are useful to the Jap as a base for operations southward and to us as a base for operations northward.

We hold the ground and continue to hold it

101

against ceaseless pressure by the Jap. In the sense that the Jap has not pushed us back and that we have denied the Jap a base there, we have won a whole series of battles for Guadalcanal thus far. But we have not yet won a victory there. For, while the Jap no longer has a base there, neither have we been able to make a base out of it. Guadalcanal is not a base at all, but a battlefield.

We are winning in the Solomons in another sense too. The Guadalcanal area has become a kind of Verdun, a maw into which the Jap is pouring his power and giving us a chance to blunt it. This is not a nickel-and-dime operation for the Jap. In the last five days here, he has lost somewhere about three hundred planes. That's not a filibustering expedition. That's a battle.

In the whole Solomons operation thus far, the Jap has lost about fifteen times as much in ships, planes, and trained men — the best men he has — as in the battle of Midway where he was denied decisively access to the Eastern Pacific. In a war against a producing unit as inferior as Japan's is believed to be, attrition is a weighty factor. The Jap is throwing planes from his factories into battle. One of the prisoners taken here recently is a twelve-year-old boy who shipped as a sailor on a patrol vessel. That might mean anything you make it mean, but it can also mean that the Jap's manpower is beginning to spread thin over the immense area he has conquered.

Intelligence here has not yet coordinated all its information and interpreted the thousands of photographs made of the five-day battle. So the extent of the damage suffered by the Jap cannot yet, as of today, be estimated with any accuracy. However, all that is information that will be given you in a communiqué from the Navy Department, which, no doubt, you will have read and forgotten before you can read this.

Jap losses were much greater than ours. There is a conclusive reason for believing this, but I am not allowed to cite it. Damage has been done to two Jap carriers, one Jap battleship, three of their heavy cruisers. Nobody has had time to count up the destroyers yet, they being only $4,000,000 or $5,000,000 engines of destruction, and therefore comparatively unimportant.

What is believed to be the best and biggest carrier the Japs have is known to have suffered four direct hits with heavy bombs. Naval gunfire, in revenge for United States Navy Torpedo Squadron 8, which sank a Jap carrier at Midway and came out with one survivor, wiped out a whole squadron of Japanese torpedo planes before one of them could get close enough to the target to release a torpedo.

103

Japanese Imperial Marines and infantry left at least two thousand of their dead behind for us to bury at Guadalcanal.

The fourth battle of the Solomons was a naval battle, but like most of the engagements in this prolonged and by all odds most curious war between navies in history, surface vessels did not figure in it vitally except as defenders of the pivots from which blows were swung. Airplanes were the striking weapons, carriers the pivots from which the strike was launched, and another of the pivots was the land on the northeast edge of Guadalcanal, where the Japs and our forces slashed and blasted at each other over treacherous terrain with tanks, artillery, and, finally, bayonets.

In the third battle of the Solomons, which ended a week ago, the Japs had forced a landing on Guadalcanal of troops and supplies. They had come in with five transports and a protecting screen of warships. Our planes had vaulted over this screen and sunk three of the transports, but at least two had got through. Then, apparently to spare their surface vessels further punishment and give their newly landed forces a chance to pull themselves together, the screen of Jap warships withdrew, one of the battleships in the force limping awkwardly.

The landing had been made very close to our positions, either because the Japs have no communications through the hundred miles of jungle and eight-thousand-foot mountains from the western end of the island (which is theirs) or because they actually intended at the time to fight from their ships as we did when we occupied the island.

After the landing, the Jap foot forces and tanks withdrew from contact with our troops and the nervous business of sending out feelers in small squads and larger patrols began and endured for exactly a week. How nervous this business was is indicated by a Marine I spoke to in the hospital here. He was on sentry duty on the beach. The duty seemed long to him, and after a few minutes of it he began to sing to keep himself company. On the third or fourth bar, as he remembers, of 'St. Louis Blues,' he was shot in the groin by a Jap sniper and had to remain silent until his relief crawled up and found him. 'If I had hollered out,' he said, 'the other fellows would have come to help me and they'd have got it.' This happened on Wednesday evening, October 21, during the 'lull' deep in American territory.

Then two powerful Jap task forces put to sea and began to converge on Guadalcanal. One force came down from the northwest, from the bases in Bougainville, Rabaul, and perhaps Yap. The other force, spread over several hundred miles, came down from the northeast, apparently from Jap bases in the Carolines, the Gilberts, and Marshalls. At the same time, the Jap land forces on Guadalcanal began to move.

The objective of this concerted move was easy to detect — the reduction of Guadalcanal. But their tactic, which had to be read by our staff officers from the flickering impressions received by aviators hurtling through the fog of war and the spearheads of weather Nature's front was throwing out, was more difficult to interpret.

This reporter has not been made privy to staff deliberations, but he saw the lights burning and the faces paling with thought. And, from a portion of the information available to the staffs and from subsequent events, a guess can be made that the Japs were intending to use their carriers to give air support to their land forces and break down our own fleet, so that their surface ships could come in close and give artillery support to the same land forces with their big guns.

Apparently, too, the task force coming down from the northwest was only making a feint, even though a very forceful one, while the task force coming down from the northeast was the business. Anyway, this was the task force we elected to hit hardest, jabbing with the left hand and throwing Sunday punches with the right hand. And our staff seems to have come to the right decision because the Japs failed of their objective.

The first intimation we here had that a full-scale battle was in the making was when the Japs stepped up the tempo of their pepperings of the area. On Friday, October 23, the air raids on the airport at Guadalcanal began to take on the lead-based complexion of a *Blitz*.

In one raid there, sixteen Jap bombers came over with a support of twenty Zeros. Grumman fighter planes (and, naturally, their youthful pilots) shot down all twenty of the Zeros and got one bomber certainly, three probably, without losing a single plane or pilot. On the same day, two Jap destroyers sneaked in close in broad daylight, threw shells at our men, and ran away safely. This need not be taken as any reflection on the alertness of our forces. Those fellows are fighting for their lives and are alert. But when fighting surface ships and bombers with fighter planes, the problem is of gasoline and when to expend it. Fighter planes carry a limited supply of gasoline and the Japs, who are no slouches at timing, got in such blows as they did when our planes were on the ground being refueled.

107

Throughout Saturday, October 24, weather was flocking in on the Battle area and piling up, and the Japs played a vast cat-and-mouse game with our forces, they hiding under the weather, we scuttling and beetling through it in an effort to locate them and determine what they were up to. The day was given over to the land fighting on Guadalcanal and to clashes between searching planes who shot each other down when cornered or when stumbling on an easy kill.

Sunday was a day which did not seem at all like Sunday here. Early in the morning, First Lieutenant Mario Sesso, flying an Army Flying Fortress (he can't decide whether to call his plane 'The Bronx Bomber' or 'The Bronx Bird') on a search mission, happened through a break in the clouds below which a sizable Jap force lay. The whole battlefield began instantly to twitch with radio messages and in a very little while to erupt with the full, awful force of war, continuing then without intermission until the Japs decided they could afford no more and withdrew.

Finding an enemy fleet is a hazardous occupation for an aviator. He not only has to find it, but stay with it while it is trying to knock him down. The PBY pilot who tipped off the Midway explosion summed up his whole enterprise by radioing his superiors the position of the Jap ships and their number and nature and direction, and then added poignantly, 'Please notify my next of kin.' Incidentally, he was overpessimistic, for he lived through his experience.

The final, most violent phase of the battle hit first at the bull shoulders of the Flying Fortress Lieutenant Sesso was flying. A swarm of Zeros came buzzing after him to shoot him down so that he could not continue to radio back the course changes the Jap ships were making to throw off pursuit.

Lieutenant Sesso played hide-and-seek with the Zeros in the clouds, but he had to come out every so often to keep an eye on the ships. The first time he came out, a Zero was right in front of his nose and facing him. His bombardier, Sergeant Eldon M. Elliott of Idalia, Colorado, flying his first combat mission and doomed to die young, fired into the Zero at the same moment the Zero fired into the Fort. The bullets crossed over each other in the air. Sergeant Elliott's bullets blew up the Zero and it shucked out its dead pilot and fell in pieces. One of the Zero's bullets hit Sergeant Elliott in the heart and he was dead before he could fall backward into his chair. In the next hour, Lieutenant Sesso saw two more Zeros fall away from him, smoking. They

109

are counted as probables because he had not the time to watch them hit the sea.

The business of finding enemy surface forces and sticking with them until your side could come up in strength and do some good went on continually during the days and nights of the fourth battle of the Solomons. The Jap forces, as they entered the arena, and ours, too, split up into something like little floating pillboxes, separated and deployed in what they hoped would be an impregnable manner. They kept shifting position continually while edging in the general direction of their objective and each new position had to be discovered and attacked.

The last covey of Jap ships was discovered at midnight Monday night by a Navy plane piloted by Lieutenant D. L. ('Jake') Jackson of Freeport, Long Island. After lingering on the scene for an hour to give his support a chance to start on up after him, Lieutenant Jackson made a torpedo run on the carrier in this force. Torpedo planes are thought to have a modest chance of survival only when a concerted attack is made on all quarters of a ship at once in combination with dive-bombers, all together dividing the fire of the ship so that the attack may be pressed home and some of the at-tackers can remain alive through it.

110

Lieutenant Jackson, a blond, soft-voiced young man with a soft skin all pinked over by youth, and his crew understood this very well, but they elected to make the run all by themselves, anyway. They had been in battle for five days by then.

Torpedo runs have to be made at suicidally low speeds because aerial torpedoes are delicate instruments and flop over when dropped too fast or from too high an altitude. Lieutenant Jackson made his run at a hundred and twenty miles an hour about thirty feet over the water. He gave the enemy a square shake at a target for a very, very long time. Every gun in the force was fired at him. The big cannons were fired into the water to throw up splashes and wreck him. His intended victim kept making violent turns to evade him. But he continued steadily through the maelstrom, dropped his torpedo when about four hundred feet away from his target, saw the torpedo land with a soft squash in the black water and proceed correctly on its way, and, as he turned, heard the noise which told him his torpedo had arrived at its proper destination.

Japanese and American search planes swarmed over the whole 250,000 square miles of battle area. They were threads, twitching and curling and connecting the various centers of battle on the sea.

In the meantime, land fighting was proceeding as

111

dolorously for the Japanese as sea fighting. Their Imperial Marines and infantry were working their way forward slowly. Small fleets of planes would dart in from the sea to help them. Occasionally, a few Jap warships would break away from the furious banging our Navy was subjecting them to and come close enough to Guadalcanal to help their troops with big guns.

Well, that, roughly and on the surface, is what the battle looked like — a corner of it given over to 'better 'ole' land fighting between men on foot or on armored wheels; relatively small patches of sea lashed volcanically by bombs, shells, and bullets; the whole stitched together by search planes, patrol boats, and submarines, and buttressed by land and sea bases for services of supply and maintenance. And into this whole churned-up segment of the world, Nature poured its own fury of winds and rains.

Only hints of the battle-rage with which our young men fought can be given by a single reporter.

The Navy vessel which took revenge for Torpedo Squadron 8 (wiped out by the Japs at Midway) was subjected at the time of its moment of triumph to a concerted attack from a squadron of dive-bombers. The Japs were out to kill. Their torpedo planes came in on all quarters and blew up in the distance like human bombs throwing themselves against

112

barbed wire. Their dive-bombers dove on all quarters. Some of these blew up in midair with their own bombs, but most plummeted blazing straight down into the sea, falling like a rain of fire all around the ship. The Japs scored no hits at all in that attack, but they came back again and again, and finally hit the ship so hard it had to be sunk.

A Navy dive-bomber squadron out to get a Jap carrier was under the desperate assaults of a great mass of Zeros for the last twenty minutes of its approach to the target. Our rear gunners shot down fifteen Zeros in this approach, killing expensively armored Japs at a rate of one every eighty seconds. The Zeros were of Varsity caliber. They were off the best existing carrier the Japs are known to have. They realized that if our squadron landed its wallop, they would die, anyway, having no place to land, so when our bombers peeled off for their dive, the Zeros followed them right on down into the withering fire from their own carrier. Our tail gunners kept shooting at Jap planes all through the dive while reporting altitude to the pilots (so that they wouldn't dive too low and hit the sea). Eighty per cent of the planes in this squadron got direct hits on the carrier. The carrier shot down more of its own planes than it did of ours, for the simple reason that there were more of theirs in the line of fire.

113

This particular attack was led by Lieutenant W. J. Widhelm, who got from his experience the longest view to date obtained by an American of a Japanese fleet after it had been struck by us.

Widhelm was shot down during his approach to the Jap carrier and was picked up at sea and returned safely here two days later. During his approach to the Jap fleet, Widhelm said that a Zero made a head-on run into him. He set his sights on the Zero's cowling, pressed the trigger and held it. The Zero ran right into Widhelm's tracer bullets, bounced straight up like a ball that is hit on the nose, and flipped over on its back, belching smoke, and fell, burning, into the sea.

Another Zero tried exactly the same attack and it was like a movie film being run over again, this Zero also bounding into the air like a ball batted on the nose and falling, burning, into the sea. Then a third Zero came along, but this one cagily looped back and forth over Widhelm's head out of range of his guns. It lingered that way for a moment, like a balked hawk hovering about, until a fourth Zero came in on Widhelm's level, then it dove straight down on him, shooting plenty of trouble into Widhelm's motor.

Widhelm, leading a formation of bombers, tried to keep in position with a broken oil line and a

114

smoking motor, but was slowed down so much that those behind him had to break up the formation to stay in back of him. Widhelm gave up in his effort to drop his bombs on the Jap carrier and tried to make his home base, but finally had to land on the sea.

He could not get his flaps down and had to make a dead-stick landing, which is very rough business, particularly for a land plane, but the plane refrained from cracking up and Lieutenant Widhelm and Radioman Stokely were able to get out a life raft, furnish it tastefully with parachute kits and practically all the comforts of home — except water. This occurred about 11 A.M. Monday. Just before hitting the water, Lieutenant Widhelm saw the Jap carrier burning, but after he had got his raft set up it had disappeared beyond the horizon.

Navy Lieutenant Atwell, piloting one of those awkward, clumsy PBY's, supposed to be useful only for search, made a dive-bombing attack on a Jap heavy cruiser at an angle of between fifty and sixty degrees. This the Navy rates as true dive-bombing, calling the kind of dives the Germans make glide-bombing.

Lieutenant Atwell, an enlisted man up from the ranks and out to show why, scored one direct hit on the cruiser just aft of the smokestack, which is the place to hit those babies and destroy them. He came so close as he dropped down on his target that the débris from the bursting ship riddled his plane, splashes from near misses filled the whole tail of his plane with water, and the concussions from his own bombs blew up a flashlight one of the crew happened to be holding in his hands. This small, strange explosion, probably caused by the air pressure inside the flashlight being greater than the air pressure outside, caused the only wounds suffered by Americans in the attack.

A striking force of Army Flying Fortresses was sent out after a heavy cruiser. They attacked in two formations, one close behind the other. As they began their bombing runs, their intended victim started turning in a 360-degree circle. They found him in a placid sea and his wake gushed out behind him like a squeal of terror from an animal running faster than its own sound.

The Jap had almost completed a full circle when the first formation dropped its bombs. The tail gunners in this formation were beginning to fill their mouths with curses because only one bomb made a direct hit and that one landed on the bow. 'I already had the words in my mouth and could taste them there,' one of them said. But before any words could be uttered, the second formation hit

116

and bombs raked the cruiser from bow to stern. The heavy ship seemed to leap up into the air and then settle into a yielding cushion of sea, belching infernos of flame from every wound. When the Fortresses landed here, about two hours later, the men went to mess with the look in their eyes of those who have seen an incredible sight.

Jap land fighting was both good and bad. Their Army was bad, their Imperial Marines very good at everything except the bayonet. During these bloody days, Japanese infantry made charges that were pathetic. The tiny figure of a Japanese officer would rise up from the torn earth far away and wave a sword. Clusters of little men, all huddled together, would rise up after him and come forward, stooping over into our machine-gun bullets and falling down dead or wounded in them.

The Imperial Marines were as good as anything our fellows have ever seen. These were the troops who broke through and got there and put up a tough fight about getting out of there. They used modern infantry tactics and filtered forward in their charges, keeping widely dispersed and flitting from cover to cover. In that way numbers of them evaded our machine-gun and rifle fire and grenades, but those who did had to face bayonets in the end and either ran away or concluded their lives that way.

117

Now that the battle is over, the situation has returned for the fourth time to its *status quo ante*. The Japs hang over Guadalcanal in the north all the way from east to west like swords of Damocles. They will range and pepper and harass and do everything they can to make the island a battlefield instead of a base.

The Tokyo Express

★

CHAPTER 8

The Tokyo Express

GUADALCANAL, SOLOMON ISLANDS

A section of what is called here 'The Tokyo Express' came chugging down from the north yesterday (November 9) and began to make a problem of itself for our forces.

This was in the cool of the evening, a very refreshing coolness indeed, and pleasant after the heat of the tropical day, and men sat around waiting for the problem to come to a head. There were no porches to sit on, only sandbags along the edges of the dugouts, but sandbags actually seemed like 'An extension of Main Street,' with men sitting there in typical porch attitudes — knees crossed, faces lifted toward the sunset, pulling leisurely on pipes.

'The Tokyo Express' presented its old familiar problem — sending in one light cruiser and ten destroyers for the sneak punch which is intended to land several thousand troops and supplies. The naval craft throw hundreds of shells and then run away.

121

Each of these night landings is a separate problem, and all together they build up into quite a splitting headache, as the Japs, when they feel that they have landed enough men, put their engines of destruction into high gear and come shooting at this little beach-head with the freshly landed troops and with substantial units of their fleet and air force.

The proper compress for these headaches is the application of all our weapons, but last night's symptoms of this woefulness, this little Jap force threading down from the north like the root of some malignant weed, was put into the lap of Navy and Marine aviators. As evening began, it was reposing there tranquilly 'in its accustomed position,' as one operations officer put it.

Official communiqués will have informed you of results long before this dispatch can, but in case you've forgotten, the Japs' effort to throw the sneak punch resulted in bangs on their own noses this time. Their cruiser and their two destroyers were hit and put out of action, if not sunk. Fifteen of the force's twenty supporting planes were shot down.

So the Japs, feeling naked without the protection of their planes and the cruiser's guns, high-tailed toward home to await another dark night for another try. Incidentally, the Japs prefer to load troops on destroyers rather than transports when operating in such comparatively small forces, because destroyers move faster, are harder to hit, and pack a wallop of their own.

This story might be called hot time on a Saturday night in Guadalcanal in which our men have lived in foxholes and in cockpits that have become so familiar the men seem to be in ruts.

At the beginning there was weather climbing down from the sky to the north. The sun was setting noisily with a splash of colors as loud as trumpet blasts, and above the blare of color lay weather, mouse-gray and soft and furry-looking at the bottom and billowing up to black at the top. The wind was making its way gently through this monsoon, bursts of howlers and gales trailing on behind it.

The weather was the main topic of conversation because the enemy force was hiding under it, and waiting American pilots would have to fight their way through it in order to hit the Japs and then fight their way home, by which time they agreed that the monsoon would really have piled up all over everything.

Blind landings by exhausted pilots whose nerves had been buttoned up in the battle were made even more unpleasant by the fact that pilots have to get down very low in order to see anything, and eight-thousand-foot mountains that are too high to vault over and too thick to brush away lie waiting in the dark toward home. Nobody considered the possibility of not having to face the problem of landing, and most seemed to regard it as the evening's major worry.

The queer antics of a monsoon in a matter of seconds can drive even a Flying Fortress (as this reporter discovered two days ago) down like a hammer hitting a pin into a pincushion and then boot it booming upwards perpendicularly. The whole fall and rise covers an area of more than a thousand feet, and all this placed second on the worry list.

The next worry came from the fact that a compass course would have to be steered home, the sky being starless, and compasses have the habit of developing unpredictable variations when jarred by machine guns. The Japs, themselves, nesting among their ferocious cluster of fire-power, placed fourth down there among such also-ran worries as where the mail from home was.

When night came down, as it does here with a pounce, rain began to fall and talk was adjourned until we filed into already sopping-wet tents. There was really no halt to the steady, quiet burble of talk. All the tents here have shrapnel holes and the men caught in the downpour would shift places as the wind shifted the rain, but they talked through that and through the sound the guns made. The Marines, eight miles east, were brewing up some powerful potion. The Japs were replying nervously, and the guns groaned, wailed, and chattered in the jungle like animals in pain. They were so close they seemed to be working inside our ears, and there were several big guns that sounded like no animal on earth. Their shells rushed past us like burning freight trains, rumbling wildly, and as if the speed of their passing had set their flames to roaring.

But the men had a trick of making themselves heard through all this uproar without even raising their voices.

Operations kept the men waiting an uneasily long time. Hitting a surface force with airplanes is a one-punch operation. And if you do not land, you generally do not get time for another wallop. In addition, you are kept struggling with a tremendous number of unpredictable things, uncertainties, and complications.

Then word came to go, and the men started running down the muddy bomb-and-shell-pocked hillock toward their planes. The blacked-out night swallowed them up very quickly.

I remember there was one young lieutenant telling

me what he was going to do on his leave on the mainland when he got it. He was going to get his wife to come out from Atlanta to meet him on the West Coast, and then they would ride back to Georgia together in a drawing room. He finished the story while still running, shouting back at me, 'I've got all kinds of money, but no place to spend it here, and I've never been in a drawing room yet.'

The pilots ran past foxholes, dugouts in which men were sleeping with their clothes on despite the rain, because everybody here knew for sure that if these boys missed, the Japs would not!

The Jap ships were spread out fanwise thirty-five miles wide when our boys found them. They had eight biplanes and twelve Zeros thrown across their prow and our fellows plowed through this welter of cannon and machine-gun fire before lowering themselves into the really heavy stuff being thrown up by the ships.

The fight blazed in the storm and darkness as briefly and hotly as a gunpowder fire, and the men, hurtling through the pelt of bullets, snatched only flash views of the action. But they had the satisfaction of seeing two torpedoes of heavy poundage hit the cruiser which was the main target, one torpedo blast one destroyer, bombs hit a second destroyer, and, as a kind of afterthought, remembered knocking down seven of the eight biplanes which got between them and the target and five of the twelve Zeros. The Zeros evidently were less impetuous about getting in the way.

As a final fillip to the action, our fellows, busting off over the horizon with all their bombs discharged, heard two tremendous explosions, apparently caused by flames reaching the cruiser's magazines.

Then the boys came home, and some of us, hearing of the good they'd done, climbed out of the dugouts and went muddily to bed, while others slept on undisturbed until morning without knowing the Japs had been crippled, and dreaming in the back of their minds of when the bombs from the Jap force would start falling.

In the course of an informal conversation here, Vice-Admiral William F. Halsey, Jr., United States Naval Commander in the South Pacific, emphasized what all who have fought the Japanese believe — that, unlike the case of Germany, there is no short cut to victory over Japan.

Against Germany, there is the possibility of underground elements coming aboveground and cooperating with an Allied military thrust, at least by embarrassing if not disrupting the home front. But American officers here, who have been investigating why the Japanese, from the merest private on up, fight as implacably as they do, have concluded that no such help may be expected from the Japanese people any time soon.

Admiral Halsey said he thought the war could not be ended short of an invasion of the Japanese mainland, and added, 'I hope so and I want to be there to see it. The man who controls the seas is the man who will win this war,' the Admiral declared, and pointed out that the seas cannot be controlled without bases and can be solidified only by sinking Jap ships. 'We have got to sink ships, sink ships, and sink more ships.'

This was interpreted as indicating that Admiral Halsey is at present convinced that island-by-island rolling back of Japanese power, coupled with a war of attrition, is the best way to conduct our operations now. Observers fresh from the United States mainland regard these as very blunt facts, indeed, but usually only one or two contacts with the Japs are necessary to convince them that the sooner Americans back home face these facts and stand up to them, the quicker the job will be done.

Marine officers here, many of them long-time residents of Japan who speak and read the language, have made painstaking efforts to discover why Japanese soldiers and their officers almost invariably are ready to die rather than surrender. The Japs express this readiness with fantastic and frequently useless acts of desperation.

In the first place, no direct evidence has been found here of a suicide impulse among the Jap troops, nor does it seem likely that the Shinto teachings, that a man dying on the battlefield goes straight to heaven, play any considerable part in dictating the action of troops in the field. The Japs are people like anybody else. While there are intense Shinto believers among them, most of them seem like most of the rest of us, and, as regards after-life, they are for it, all right, but meanwhile they are willing to be content with a bird in the hand rather than in the bush.

Secondly, among more than six hundred prisoners taken in the operations here so far, most singly or in small batches, not one is an officer. Prisoners have been taken only when separated from their units.

Thirdly, the evidence indicates that the Jap soldier, when beyond the personal observation of officers, is definitely reluctant to give battle unless he is pretty sure the odds are with him. This is apparent in the case of Zero planes manned by noncommissioned officers or privates or in the case of isolated machine-gun nests.

Fourthly, and perhaps most provocative, Jap sol-

129

diers in direct contact with their officers keep trying to kill our men even in the last extremities and even after having suffered torments in the jungle calculated to rob the average mentality of all will except the desire to survive.

Finally, no Jap officer has yet been discovered who was reluctant to throw himself and his men against any odds or was minus a willingness to die rather than surrender. That is so basic that it seems deeper than habit or instinct and to be almost a reflex.

Opinions derived from these facts make a long story, but briefly they are that Japan's ruling class, from which the officers stem, had the firmest control over its people; that Jap troops are convinced it will be impossible for them to return home if they are disgraced in battle by showing weakness before the enemy or by being taken prisoner; and that victory for us will be difficult until enough officers have been killed to impair their control over troops and sailors. The Japs' costly venture in China does not seem to have impaired the officers' control over the troops here, many of whom are veterans of this campaign.

A morsel of comfort in the situation is the fact that the supply of Jap officers is limited, as indicated by the increasing number of bomber and fighter pilots taken from the uneducated classes without being given commissions. It seems to be true that the feudal system, which was all right in the days when a war was decided by knights in comparatively individual combat, does not rest on a broad enough base to stand up against the voluminously talented democratic fire-power.

These opinions, except perhaps for the comforting one, Admiral Halsey indicated he shares.

CHAPTER 9

Jungle Battle

★

CHAPTER 9

★

Jungle Battle

*WITH THE UNITED STATES FORCES IN THE
JUNGLES OF GUADALCANAL*

Veterans of Central America bushfighting and
of Belleau Wood and the Argonne who are now here
describe the terrain over which our forces have been
advancing yard by blood-stained yard since November 1 as the worst battle terrain any army ever
fought over. And this is probably underestimating
the situation.

Japanese habitués of this terrain, most of them
veterans of the Malay and Java jungles, tried very
hard for about five hours today to get this reporter.
On any ordinary battlefield this would have been
like potting an overstuffed pillow in the parlor for
any trained soldier, your correspondent having the
bulk of the average general without comparable
agility in concealing same.

The Japs tried first with snipers, then with machine guns, and finally brought up mortars. I suspended breathing frequently, but never permanently, and the closest the Japs could come was to bedeck me with clipped-off leaves of the local vegetation and to freshen my complexion with mud-packs applied in clumps from the rifle bullets that smacked into the earth around me. Even new Marines coming up to the firing line for the first time, and thus still respectful of the invisible terrors of the jungle (it takes an average of two days for the green troops to learn their way around), opened up on me for about forty minutes without damaging anything except my nervous system. This proves how exasperating this terrain is for an army trying to get somewhere.

The tactics of the Jap here are exactly the same as he used in his *blitz* through four thousand miles of Allied bastions for the five months ending early last May. But the Marine tactics which the Army is faithfully following here blows the Jap *blitz* right back into their faces. And now, since we started our first large-scale, full-panoplied land offensive on November 1, we have proved capable of generating plenty in the way of a 'Made in the U.S.A.' *blitz*.

The Japs have tried to dam up the slow, steady, and paralyzing flow of our green-clad troops into their territory here with everything they have, including landings of new forces to our rear, but the results to date have been all in our favor. The main offensive westward of the air fields advanced about three miles beyond the Matanikau River in nine days through terrain where a man, walking without opposition and without cares, would be doing well to average two hundred yards per hour, or about one mile in nine hours.

The terrain eastward does not feature the fortress-like ridges, honeycombed with caves and matted down with unmanicured jungle that make the western front as mortal as a beehive. Thus, a regiment of Japs, landed on the west off destroyers early on November 3 in an effort to force us to thin out our thrust, was able to lead us five miles down the coast before getting themselves bottled up. As this is being written, this particular force has water at its back, and an infantry unit and Marines under Brigadier General E. B. Sebree of the Army has them ringed around from beach to beach in a semi-circle with a radius of six hundred yards, and automatic weapons are pouring on a real Corregidor shellacking.

There is nothing new in our tactics here, according to General A. A. Vandergrift. 'Hell,' he told this reporter, 'I've been fighting this type of war the same way since I was a kid of a second lieutenant.' British officers, in Malaya, facing unknown and undetectable fire power on the flanks and in the rear, had the habit of fighting their way backwards to conserve their force from what they believed a formidable 'Kiel und Kessel' Tokyo style, but we simply stand our ground, wait and see.

We are daytime fighters, and when twilight comes, we revert to our Indian-fighting past and build old-fashioned squares of defense around each separate automatic weapon. In the morning we resume our advance, leaving rear echelons to destroy whatever has filtered through. It was while going up toward the front lines that this reporter was fired at, and got a case of squeaky nerves.

All these goings-on would fill any military map of so-called fronts with very queasy lines, which are hardly lines at all, but really just X's to mark the spots where the bodies lie.

It was a forty-five-minute drive in a jeep from the tent where the press was quartered to where the two-hundred-yard deep western front begins on this embattled island. Our boys are already swimming and washing clothes in the Matanikau River this morning while the rumble of guns on ahead is making the water quiver with a massaging effect.

'I figure,' said one expert launderer, 'three more batteries are needed to turn this river into a real, high-grade washing machine.'

From the river on, the jungle shore is withered and blistered with blackened clumps where the Japs held their line with what amounted to natural pillboxes. Artillery dug them out or plowed them under. The smell of dead bodies still hangs, in clouds and thick as smoke over these narrowly separated places where men who had traveled four thousand miles elected to stand and die. Huge trees which were felled or severed by shells are sticking up like gaunt, amputated limbs, like the wounded of the last war in Vienna who had turned beggars and used mutely to hold up their unbandaged stumps to passers-by.

Our fellows have elected to die here, too. Along the road pounded through by the wheels of our trucks lie several graves, one of a private. His friends have trimmed its mound pathetically with coconuts and fashioned a rude wooden cross for a headstone. A helmet with three holes in it, the holes as blank as dead eyes, tops the cross and on it is penciled, 'A Real Guy.' Against the cross stands a photograph of a very pretty girl, staring silently. The sunlight is very bright here and you can see the brown color of the eyes in the photograph and seem to be able to look deep into them down past the look she gave the Dodge City, Kansas, photographer and his camera. Her dead man must

139

have loved her truly, for he carried her picture into battle.

The flat shoreline runs only a short distance along the sea and then begins precipitously to climb up into ridges which mount rapidly into an eight-thousand-foot mountain range. Each ridge drops steeply into a ravine, known as a draw. The ridges are about five hundred yards apart and run from north to south. Since our advance is east to west, our troops have to go up and down steadily as if over waves. The jungle runs up along the sides of each ridge almost to the top, and then on the top there is bare grass, completely bare, not a tree or log or twig. The Japs stay down in the ravines and fire up at the bare space on top. That means our fellows going up and over after them are clearly outlined while the Japs have dense, actually impenetrable cover. Complicating the tactical problem even more is the fact that the jungle-covered sides of each ridge are pocked with naturally eroded coral caves which the Japs have enlarged and in which they entrench themselves obstinately. Flame-throwers are no good against these caves. The jungle, through which sunlight never penetrates and in which only lizards and red orchids flourish, is too damp to permit flame-throwers to function. The Japs either have to be interred in their caves or excavated by hand.

140

American troops, no doubt, will face more formidable opposition before this war is concluded, but never a more desperate terrain filled with more desperate men.

On the eastern front, tanks are used to flush the Japs out of the grass, and when they are flushed, they are shot down like running quail. But tanks are useless on the western side of Henderson Field. On the western side, it is all manual labor.

We sat on a ridge which had been won yesterday, then squatted to watch our Marines move up to the next ridge about five hundred yards westward and start scooping out the succeeding ravine.

The second wave of Marines worked their way forward slowly, flowing up the covered side of the ridge in the slow, purposeful uncertain-seeming way of raindrops on a window, stretcher-bearers coming slowly down among them from the top. There was a mingling of Marines and corps men, and suddenly, in the midst of the mingle, flanking mortar fire dropped. Wham! The crack of it came back to us hard enough to sound like a hammer against the flesh of the brain. Then, wham again and wham! Three fifteen-pound shells in all, and the jungle's trees, all gnarled and burdened with parasitic creepers, bent and thrashed and groaned under the blasts. The Japs sent over only the three, evidently

141

being short of heavy ammunition which they had been unable to drag with them in their retreat.

I think I closed my eyes when the first shell hit. It seemed too horrible, our men and the stretcher cases forming a cluster to cheer each other up and then becoming a bull's-eye for the Japs. When I opened my eyes, the ground was bare. There were three holes in it, very close together. I thought everybody there had been killed, and I sat there a moment thinking that. Then Marines began to lift themselves from the earth to which they had dropped, and one by one, heads down, gathering up their rifles and gear as they went, resumed their slow, purposeful, upward flow. They never looked back. They just went on up and over.

Then down the other side, cautiously, probing, seeking to draw fire in order to locate Japs to shoot at. They drew fire all right — three machine-gun nests clustered so close together in the jungle that a grenade attack would most likely have been sufficient to wipe it out. But the Marines now enjoy the luxury of artillery and nobody is wasting men around here — not our boys, not the kind of fellows we have here. So everybody waited for the big boys to go to work. They waited patiently, dug in one hundred yards from the nests.

It seemed as if it might be a long wait, and we

scuttled through the sniper fire to the command post where Colonel John M. Arthur, of Union, South Carolina, directing operations on the whole front, Lieutenant Colonel Cornelius P. Van Ness and the artilleryman, Colonel William Keating, of Philadelphia, were ensconced. Fifteen Zeros, one of them burning, were working up some sort of hell in back of us back near the airport, at the same time that rifle and machine-gun fire was cracking all around us, and officers were trying to make themselves heard on the telephone above the din.

Captain Maxie Williams, a blondy, chunky, footballish young man, standing recklessly in the open trying to isolate the Jap harassing fire so it could be smothered and the command could be left to command in peace, called for three volunteers to swing around and work over to a certain tall tree. Sergeant Major Frank Regan, of Dallas, Texas, and combat reporter Sergeant Ned Burman, of San Francisco, crawled forth, Burman observing sallowly, 'I forgot the first rule of the Marines — never volunteer for nothing.'

I remarked to Captain Maxie that all this going-on reminded me of Harlan County, Kentucky, on election day, where politics are so mixed up that you cannot tell whether you are shooting a friend or a foe. And Captain Maxie leaned into the crater

143

where we were hiding from the bullets to shout, 'Sure ain't like Humphries County,' he being from Waverly, Tennessee.

Through all the shooting, Colonel Keating directed the artillery fire toward the Jap machine guns holding up our advance. He worked like a surgeon probing for an ulcer. The first thing I heard him say — he had to shout over the rifle and machine-gun fire — was, 'Now, just be calm and careful. We don't expect you to do any better than your best under the existing circumstances.' He was talking to a boy in a forward artillery observation post whose job it was to direct the artillery fire. The boy, being so close to the target, was nervous.

He gave a target and Colonel Keating ordered one shell fired. We could hear this boom! like some brass-lunged giant clearing his throat, and then the feathered whistle of the shell. 'It's passing right overhead now,' said Colonel Keating. 'Just watch for it. It ought to be there any minute.' Then we heard the boom of the explosion.

Then the boy gave another target. Altogether, three ranging shots were fired. The fourth shot was right on the target. Colonel Keating asked the same battery to fire a salvo of five shells, and we heard the booms all run together and the feathered whistles all run together and the explosions like

144

one long explosion. Then silence. 'Do you want any more?' Colonel Keating asked the boy, and I could hear the boy's voice squeaking exultantly through the telephone. 'That's enough, sir,' he said.

All advances here are like that, inch by inch, slow and easy, and the Japs don't seem to have anything to stop them except another one of their combined-operations attacks — a fleet by sea, a fleet by air, forcing us to fall back on the airport and bury there the hell the Japs raise.

The Fifth Battle
of the Solomons

★

CHAPTER 10

★

The Fifth Battle of the Solomons

FROM GUADALCANAL

The fifth battle of the Solomons, which in many ways proved a Japanese disaster unprecedented in the history of the world's great navies, began with a dispute over reinforcements of men and supplies for our embattled land forces on Guadalcanal.

Reconnaissance had revealed that the Japanese were building up an extensive force to retake the Solomons, but we threw the first punch, landing the initial wave of our reinforcements on November 11. We held the initiative that day and on November 12 when the second wave landed.

The reinforcements were sufficient to make General Vandergrift of the Marine Corps remark to this reporter, 'I now feel it is no longer possible for the Japs to land enough strength at any one time to take Guadalcanal away from us.'

So the Japs had quite a target to shoot at, and they shot and shot again, and shot four times altogether, missing each time. As he was shooting, the Jap was hastily forming convoys behind his first line of fire, an invasion force estimated by Intelligence here to be three divisions with full equipment. The sea train formed consisted of eight transports, one an NYK liner which is the biggest the Japs have. The other seven ranged downward from eighteen thousand tons. Accompanying these troopships were four cargo vessels of ten to twelve thousand tons each, carrying the raiment of the Emperor's divisions, while screening the array were at least four battleships, plus a complement of cruisers and destroyers.

On Friday, the thirteenth, this force, making up the bulk if not all of Japan's South Pacific Fleet, moved into the arena and wrested the initiative from us. This transformed the character of the dispute into a replica, only more so, of all the previous Solomons battles — a skillful, tenacious, heedlessly bloody attempt to reduce permanently our Guadalcanal salient by wiping out its sea armor and obliterating its garrison of planes and men.

At 1:40 on Friday morning, United States forces here began fighting, not for some future offensive with the Japs, but for their lives. The result of this desperate, completely reckless fight for life by the Americans was twenty-eight Jap ships sunk, including two battleships, plus ten damaged. This

150

reporter, being less conservative than the United States Navy and more willing to trust the evidence of his own eyes, personally scores more than half of the ten 'damaged' as being sunk. Our cost was seven destroyers and two light cruisers.

This unprecedented battle had many curious features which no doubt will be debated in naval academies for many years. Naval vessels fought in the night, airplanes fought in the day, both in the same arena. To climax the battle, on the afternoon of November 14 Jap warships fled from the transport and cargo vessels they were supposed to guard, and left the Emperor's three divisions naked to the assaults of our planes which were based twenty minutes away.

These were all very novel tactics, but even more novel, at least from the point of view of this eyewitness whose life was one of those immediately being fought for, was the fact that all the major actions of the battle took place within clear view of the naked eye. This is the first battle in the history of modern war that could be viewed almost in its entirety by a single man standing still.

The holocaust was unfolded to this reporter's front-row seat like a classic Oriental drama where the spectators know each move of the actors in advance through having seen the show several times before, and yet await the acts of execution with passionate and terrible anticipation. Thousands of land-bound Marines, Army, and Navy men had seats so close to the drama that the actors literally spilled from the stage into their laps. Then the spectators joined in the action savagely.

The first fiddling notes of the drama, a kind of overture, were heard on Wednesday the tenth, when a force of fifteen Zeros and bombers reconnoitered over the airport and harbor. Apparently the Japs knew something was cooking which was more than most here knew. The Japs flew high and fast and stayed far away. The inclusion of bombers was shrewd tactics, as it made our pilots take the defensive around probable targets and linger there, so the Japs did not get hit until the tailend of their force was scampering over the horizon, one Zero being snipped off like the last rattler of a snake's tail.

At dawn on Armistice Day the first wave of our reinforcements steamed into the harbor, and all ashore knew the fat was on the fire. Then it happened that the Marines' and Army's hearts became the first to break of all the hundreds of thousands broken these last four days. Our troops, who had been advancing westward steadily since November 1, scooping the Japs out of the jungle, ravines, and caves, now could advance no farther, but had to hold their lines to await the expected Jap thrust from the sea.

Some of our men were about to pounce on a cluster of four machine-gun nests when the order came to let those Japs live a little longer, and the puzzled, exhausted, and battle-soiled men furnished an emotional problem for the officers. Exasperation came to a boiling point when Jap shelling killed five officers with a single shot. Surviving officers had to plead with their men to store up their anger for a more propitious day.

In the meantime the Japs began to reply to our reinforcing move. As in the fourth battle of the Solomons, both sides were abused by the weather. In a twenty-five-mile scout I noted five kinds of weather, ranging from blue all ablaze with sun to a black thunderhead sitting on the water and stretching up as far as the eye could see.

At 9:30 on Armistice Day morning nine Jap dive-bombers guarded by twelve Zeros dropped on our transports from the gray scud along the edge of a squall. Their stroke split the eyes of the spectators on the beach like lightning, and was over long before the shock and glare of it subsided.

A new Marine squadron sent up to sharpen their spurs on the Japs suffered as must all beginners in the art of war. They lost six planes, but two pilots were saved, coming down on the water with rescue boats waiting under them. Another was seen parachuting down behind the Jap lines where he had a fighting chance of getting home. As new as they were, these boys gave more than they took — they got one dive-bomber and four Zeros certainly, and one more bomber and two Zeros probably.

But what is more important they achieved the major objective of so chivvying and badgering the Jap assault that it failed in its purpose. One light bomb warped the hatch of a nearly unloaded freighter and a near miss damaged the electric wiring on the stern of a second, but that is all the Japs achieved that time.

A little past eleven o'clock of the same morning, the time the civilized world devoted to silence in honor of the last war's dead, the Japs stepped up the tempo of the action. They sent in twenty-five heavy bombers with Zero escort. The Japs stayed up higher than twenty-five thousand feet and clung tenaciously to formation while our green fighters, learning fast and learning the hard way, stung and bit at their metallic sides, and could be seen landing there as delicately as mosquitoes on gleaming flesh.

Bombs fell among newly withdrawn Marines near the beach, killing one and wounding a few. Before the clouds swallowed up the show, forty-nine Japs, enshrouded in seven bombers, could be seen struggling for their lives. One bomber fell like a silver spear from the top of the sky. One of our fighters followed him a little way down, spitting into his wounds with machine guns, then seemed to think that enough is enough, and turned back to get himself another. But not more than two hundred feet above the water these seven struggling Japs managed to level off, and it seemed like a suddenly halted movie in which a suicide leap halts in midair.

The plane started to streak toward the clouds, but it had no friends to cheer it on. Men trying to staunch the blood of the wounded near me lifted their eyes from their task and watched the lonely struggle of the Japs for life. A minute went by and then another — a long time to those watching — and then tail smoke blew out of the Jap as suddenly as a gasp. All on the beach cheered and were still cheering when the bomber nosed gently toward the water to become a funeral pyre for the crew of seven.

We lost one fighter in that brawl, but up in the clouds got seven bombers, left one more and one Zero smoking, and again prevented the Japs from even pinking their objective. It was our Armistice Day, overwhelmingly.

155

On November 12 more of our transports arrived under convoy. There was considerable tension among the spectators, but considerable tranquillity too. At least four crap games were proceeding on Army blankets on the beach while awaiting the Japs, but photographers refused to wander from their cameras, which were set up and focused, and gunners refused to turn from their loaded guns. All ate their noon chow on the spot.

The morning's only excitement came when a Jap six-inch land gun, manned by a crew with more courage than brains, fired a single shot, missing a transport by five hundred yards. Admiral Callaghan's cruisers and destroyers and General Vandergrift's land guns fell on the Jap smotheringly and they never even squeaked again.

At 2:20 in the afternoon the Japs attacked with torpedo bombers with a Zero escort. Nobody is sure how many bombers, but it was somewhere between twenty and twenty-five. It is known that the total Jap force numbered thirty-three planes. One Zero escaped our fire, and in destroying the thirty-two others we did not lose a single plane or pilot.

The torpedo bombers launched themselves from a huge bank of black clouds into point-blank fire from our ships and enfilading fire from the beach. Only four torpedoes were seen to drop, but not one hit anything except empty water. The Japs raced toward the ships so low and so close to the beach that it seemed possible to reach out and knock them down with your fists.

The spectators on the beach fired everything they could reach, including revolvers. Six Jap ships were seen to blow up, the débris showering the beach. One hapless Jap, seeing the death of his comrades, lost heart and circled helplessly, seeking a way to escape until a Grumman Wildcat ended his misery.

The cruiser, *San Francisco*, dodging a torpedo, got a bomber which blew up over the deck, killing eighteen men and burning several others. Our fighters ran around the Ack-Ack fire, picked the Japs up on the other side, and chopped down on them from so close above that it looked like knives hacking at a butcher's block.

Only three of about one hundred highly trained Japs survived the attack, and two of these are now wounded prisoners. A rescue boat threw them a rope, but when they reached for it a Jap officer with them slapped down their hands. A high-squealing argument started and the rope was thrown again, and again the Jap officer slapped down their hands and turned his back on their protests disdainfully. They shot the officer in the back of the head and caught the rope the third time it was thrown.

157

The intensity of the action was typified by the case of Captain Joe Foss, a Marine ace from South Dakota, who raised his score to twenty-one planes in the course of the day. Once he shot down a Zero at twenty-nine hundred feet, dove down to less than three thousand feet, and chopped down two torpedo bombers. He dove so rapidly that the wind stripped the glass shelter from the hatch of his plane and all the rubber pads, and he was working on his third victim before the first completed its fall into the sea.

The whole show lasted less than ten minutes, and small boats spent until dark combing the wreckage of millions of dollars spent in Jap production and looking for ninety-four Jap bodies.

As night fell, the planes departed from the stage, and the ships with their big guns entered. Search planes had been watching the Japs all day long and their next move was expected. Admiral Callaghan's force shepherded transports to safety. The Japs apparently watched them, but lost them in the darkness, for during the night they came stepping breezily to deliver what might have been the decisive blow of the battle.

The land forces had girded themselves for a repetition of the October 13 bombardment. Men huddled in foxholes, and asked each other silently with

158

their embittered faces, 'Where's our Navy?' and wondered what would be left to stop the Jap transports.

Those seven hours of darkness, with each moment as silent as held breath, were the blackest our troops have faced since Bataan, but at the end of them our Navy was there, incredibly, like a Tom Mix of old, like the hero of some antique melodrama. It turned the tide of the whole battle by throwing its steel and flesh into the breach against what may be the heaviest Jap force yet engaged by surface ships in this war.

Again the beach had a front-row seat for the devastating action. Admiral Callaghan's force steaming in line dove headlong into a vastly more powerful Jap fleet which was swinging around tiny Savo Island with guns set for point-blank blasting of Guadalcanal, and loaded with high-explosive shells instead of armor-piercing shells. Matching cruisers and destroyers against battleships is like putting a good bantamweight against a good heavyweight, but the Japs unquestionably were caught with their kimonos down around their ankles. They could have stayed out of range and knocked out our ships with impunity, and then finished us on the ground at their leisure.

We opened fire first. The Jap ships, steaming full speed, were on us, over us, and all around us in the first minute. Torpedoes need several hundred feet in order to arm themselves with their propellers. Our destroyers discharged torpedoes from such close range that they could not wind up enough to explode. The range was so close that the Japs could not depress their guns enough to fire at the waterline, which is why so many hits landed on the bridge and two of our admirals were killed.

The action was illuminated in brief, blinding flashes by Jap searchlights which were shot out as soon as they were turned on, by muzzle flashes from big guns, by fantastic streams of tracers, and by huge orange-colored explosions as two Jap destroyers and one of our destroyers blew up within seconds of one another. Two Jap planes, which were overhead intending to drop flares on the target, were caught and blown to bits.

In the glare of three exploding ships, the two naval forces could be seen laboring and wallowing in their recoils, throwing up waves in the ordinarily lakelike harbor which hit like rocks against the beach. The sands of the beach were shuddering so much from gunfire that they made the men standing there quiver and tingle from head to foot.

From the beach it resembled a door to hell opening and closing, opening and closing, over and over. The unholy show took place in the area immediately this side of Savo Island. Our ships, in a line of about three thousand yards, steamed into a circle of Jap ships which opened at the eastern end like a mouth gaping with surprise. They ran, dodged, and reversed their field, twisted, lurched, and lunged, but progressed generally along the inside of the lip of the Japs.

Since the Jap circle was much bigger than our line, the Jap ships, first at one end and then the other, fired across the empty space into one another. It took about thirty minutes for our ships to complete the tour of the circle and by the end of the tour the Jap ships had ceased to exist as an effective force.

The whole thing was like a huge ring around a thin finger with the finger trying to burst through the ring and the ring trying to break every bone in the finger. Then the battling ring and the embattled finger seemed to crawl slowly toward us, still locked in a deathlike embrace and swayed, thundering and shuddering backward and forward.

After thirty minutes the Japs crawled out of the harbor without having dropped a single shell on Guadalcanal, but in the morning twenty new Jap landing boats were seen on their portion of the beach, so a landing must have been made under the cover of the battle.

The exact composition of the Jap force probably will remain unknown until we break open Tokyo's archives after the war. From the beach twenty-six Jap silhouettes were counted, but they were shifting shapes illuminated fitfully and duplication in counting was possible. Our force was composed of eight destroyers, two heavy cruisers, and three light cruisers, but there's not a man living who could remain a statistician before so gruesome and incalculably costly a spectacle.

The Japs had at least one battleship of the Kongo class. That is certain, for it figured all the next day in one of the most fantastic episodes of this war — 'the episode of the unsinkable battleship.' There likely was another battleship sunk. Several of our destroyers' survivors recall seeing the whole bridge of a battleship leap into the air, but when daylight came, the 'unsinkable battleship' was seen to have its bridge undamaged.

Also, flames from one of our burning destroyers illuminated a ship which was upside down with only the hull showing above water. The hull was huge, but the Navy refuses to concede it was a battleship unless concrete evidence is produced, which is impossible in the deep water where the ship sank. It is certain that few Jap ships could remain unhit in that avalanche of fire, or they would not have broken off action against a force stripped down to the bone, and with that bone broken in several places.

Admiral Norman Scott was killed a minute and ten seconds after the battle started, standing with his glasses glued on the enemy. Admiral Callaghan and Captain Cassin Young, commanding his ship, were killed about twenty seconds later, leaving Lieutenant Commander Bruce McCandless senior officer aboard the cruiser flagship. Lieutenant Commander McCandless was unable to inform the remainder of the force of Admiral Callaghan's death, so he took command of the flagship. In the next few moments five high-explosive shells landed, one at a time, exactly where he had been standing only seconds before. His valor earned for him promotion and recommendation for the Medal of Honor.

Four of our destroyers and at least two Jap destroyers and this 'whatzit,' which most of us call a battleship, but which the Navy called 'heavy cruiser or battleship,' sank within thirty minutes. One of our mortally hit light cruisers remained on the scene with a crippled heavy cruiser fumbling and floundering for such Jap ships as were unable to withdraw. They found one cruiser at dawn, and the heavy cruiser shot it to its death, turning it bottom-side up with the first salvo.

With the coming of dawn a large mushrooming column of black smoke could be seen from the beach about five miles beyond Savo Island, and the episode of the 'unsinkable battleship' took the center of the stage. A muted accompaniment was supplied by a flotilla of rescue boats, putt-putting after survivors.

If the night scene was incredible, the day scene was even more so, if in a more subdued way. The waters spread out in front of Lunga Beach like a basin filled with a bloody gruel simmering quietly in the tropical sun. In the middle of it stuck up the black lump of Savo Island, and beyond it and a little to the right was a column of black smoke.

That was the setting, and in it, as always happens when armed units are burst open into their surviving human components, thousands of separate tragedies and hundreds of little comedies with unexpectedly happy endings played themselves out all day long.

Rescue boats picked up more than eight hundred of our men, about two hundred and fifty of whom were wounded. Those not hurt were laughing and joking as they stepped ashore after hours in the oil-soothed waters.

Their most frequent comment was, 'You can't fight battleships with tin cans,' just as if they hadn't done just that.

Hundreds of Japanese sailors, small dark figures bobbing in the water, tried to continue the battle against their rescuers by shooting at all the on-coming boats. In the end they killed themselves or deliberately dove under water, staying there until drowned. The rescuers had to arm themselves with machine guns. There were no Japanese sailors with lifebelts in this battle. Altogether, only twenty-five allowed themselves to be rescued by us.

In the midst of this basin which crackled with the fire of small guns, schools of sharks threaded their way, hacking at corpses and the wounded. Over it roared steadily an all-day-long shuttle service of airplanes, running to sink the Jap 'unsinkable battleship.'

The battleship, making about five knots, was screened by five destroyers which were left behind from the battle, while the rest of the Jap force, if any, scuttled off. Captain George Dooley, of Hopland, California, led Marine torpedo planes in the first attack, and made another an hour later, scoring direct hits in each attack.

Then Lieutenant Albert D. Coffin, of Indianapolis, leading a squadron of Navy torpedo planes to reinforce Guadalcanal against the Jap transports which then were moving out of range to the north-

165

west, tripped over the spectacle of the battleship. Lieutenant Coffin paused, fascinated, and dropped three torpedoes into the ship before continuing on his way.

Then Omaha-born Lieutenant Harold 'Swede' Larsen joined the fray with his friend, Major Joe Sailer, of Philadelphia, who led dive-bombers. Major Sailer, thousands of feet above Lieutenant Larsen, synchronized watches with his friend and ticked off the seconds by radio, saying, 'Mark one, Mark two, Mark three, Mark four, and ... and ... and go!'

Major Sailer dropped as Lieutenant Larsen launched Torpedo Squadron Eight out of a bordering squall. The Navy observer reported seeing Lieutenant Larsen's torpedoes hit the side of the battleship directly under Major Sailer's bomb. This was pool-shark shooting, to say the least.

The assaults continued interminably and so frequently that the battleship, although able to go from four to six knots, could make only ten miles in the entire day.

Lieutenant Coffin's men, fresh from Navy luxury, arrived all spruced up — freshly bathed, shaved, and combed, in neatly pressed flying suits. As the endless day wore on, they gradually assumed a grimy look like the rest of us here.

Bullet-holes sprouted in ever-increasing numbers in their fuselages as the same planes returned again and again for attack. But still the battleship remained unsunk.

'We've got to sink it,' said Lieutenant Coffin, 'or else the admirals will stop building carriers and start building battleships over again.'

When night shielded the battleship from further attack, the ship's whole stern was cherry-red from internal fires. It lay bloodstained on the darkening waters, but one battery just forward of midships kept their guns firing.

The Jap destroyers and battleship had shot all day long without getting a single one of our planes and wounded only one man. In the end the 'unsinkable battleship' had eleven torpedoes in it, and four heavy bombs and three medium bombs on it from above.

In the meantime, Jap transports were milling around out of our range, 'beating their gums,' as the Navy says. Under cover of darkness a Jap battleship force plunged on ahead of them to try another attack on Guadalcanal. They reached us at two o'clock in the morning of November 14 and shelled us with salvo after salvo of six-, eight-, and fourteen-inch shells for forty minutes while Jap destroyers unloaded the 'unsinkable battleship' and scuttled it.

167

The bombardment got only three of our planes, but damaged seventeen others, which were repaired and back in the air by nightfall. The Jap force obviously was jittery and expecting the worst from every corner of the unfathomable darkness, for when they were attacked by a small force of mosquito boats they cleared out instantly. They fled like an elephant before a champing, whiskered mouse, but not before the mosquito boats had embedded a torpedo in one of the cruisers.

Showing how one arm helps another in this kind of warfare, this single torpedo cost the Japs two cruisers. The Japs had left a cruiser and four destroyers behind to help their cripples home. Our flyers, helpless in the darkness under the above-mentioned shelling, stored up their anger for the morning. Shortly after dawn, they found the cruisers and finished off both with three torpedoes and two heavy bombs.

Then a flying fortress, piloted by Captain J. E. Joham, of Santa Barbara, California, found the Jap sea train of transports, and all the fooling around with cruisers and destroyers and such was stopped for the grim business of stripping the Emperor of his divisions and their arms.

The transports were found only a hundred and fifty miles west of Guadalcanal, and again a shuttle service was run from Henderson Field. Captain Joham reported a carrier in the force which sent up three Messerschmidt 109's and two Zeros. Before he finished knocking them down, the carrier, if present, had run away, the Japs apparently being very short of carriers and more anxious to preserve them than troops. About three hours after the start of our attacks on the transports all the Jap warships ran away from the scene, leaving the troops without any umbrella against the rain of bombs.

When this happened, it caused our command to think that the Japs were up to some ruse. It seemed inexplicable to our kind of fighters to leave their own men to be massacred helplessly. But it was not a ruse.

Our flyers were sickened at the slaughter of the impotent Japs below them, but this was all in a day's work which had to be done. The Japs do not surrender, but have to be killed. The Jap leaders threw their men into a rapidly working blast furnace and insisted that they keep trying to crawl through it — and their men kept trying. Not once did they halt their feeble, floundering crawl through the hot coals being poured on them. By nightfall they had succeeded in making seventy miles progress toward us.

All of our air forces in this area joined in what had

169

become pickaxe work. The pilots, setting their faces, called themselves the 'buzzard brigade,' but kept at it as relentlessly as the Japs.

By nightfall all eight of the Japs' transports had been sunk and only four of their medium-sized cargo vessels remained afloat, two of these hit and burning. Our total cost for the day was four planes — all shot down in the last attack, possibly because the pilots were exhausted.

In mid-afternoon of the same day the Japs on Guadalcanal, informed of the disaster overtaking their reinforcements, managed to haul up a six-inch gun within range of Henderson Airport, and made a pathetic effort to interrupt the work of our planes. Of the first twelve Jap shells, nine were duds and three made new bouncing spots for our roaring jeeps.

Two hours later, the Japs fired nine more shells. Eight of these were duds and the other one plowed up a new victory garden for us in the grass. And this was the sole 'protection' the Emperor could give his three divisions.

At 11:20 on Saturday night, the fourteenth, the Emperor's divisions, still flowing toward us monstrously, like some amputated torso gushing blood from almost every inch, made a last desperate effort to take Guadalcanal. Their warships returned

170

to help them, but once again our Navy outwitted the Jap and anticipated his every move.

This time we also outgunned him, and the Jap's last effort was his costliest.

The Japs came sweeping down from the west. Our battleship force tailed them, but let them go around the north side of Savo Island while we, lagging a little behind, came around the south side, and caught them in that dream maneuver of all naval warriors — 'crossing the T.'

This battle was also visible from the beach. It is believed here to be the first naval battle in history in which sixteen-inch guns were used against vessels. It was even more spectacular and terrible than Friday morning's battle, and again lasted about thirty minutes.

In that time I counted eleven ships burning and exploding and sinking, two of them ours, but the Navy believes in erring on the side of caution.

At 11:50 the surviving Jap ships began steaming westward, firing over their shoulders. Our force gave chase, but the Japs were lost in the darkness shortly after one o'clock. But at five o'clock on Sunday morning flashes from explosions could be seen from Russell Island, forty miles away, which may have been the jittery Japs tumbling over one another and firing away at each other. The glares were so vast that they lit up the tired faces of the beach spectators even from that distance.

The morning's first light found the remains of the Jap sea train beached less than seven miles from where this is being written — four cargo vessels. Destroyers have been shelling them, and our flyers, still calling themselves the 'buzzard brigade,' have been over them without intermission, strafing and bombing.

All four ships were gutted before noon. The ship nearest us, beached by the Poha River, had no superstructure or deck, and the flyers could see into the bowels of it, which were as red as an excavated heart. But the relentless Japs still had gunners — standing in the water on the stern, throwing rocks.

The Japs managed to throw off what stores they could onto the beach, but they could not drag them toward the trees. Our flyers keep pounding the Jap corpse, cremating it with Molotov cocktails. The Jap stores are now making a fire a thousand yards long and two hundred yards wide on the beach, and we intend to keep that fire burning until there's nothing left to burn. It warms our hearts.

CHAPTER 11

The Sinking of the

President Coolidge

★

CHAPTER 11

★

The Sinking of the President Coolidge

SOMEWHERE ON THE PACIFIC

The mine that hit the liner *President Coolidge* made hardly any sound at all and was just felt as a jar by the men on board, a deep, shaking jar.

This reporter happened to be flying over the ship in a bomber at the time. The slate-gray liner was making its way slowly through a slate-gray morning. Then it seemed to lurch like a man hit heavily, and a big, gaseous-looking bubble welled up on its starboard side just about midships and hung swollen on the surface of the sea for a moment before bursting.

The ship swung hard to port, and then suddenly it seemed to lurch again and another gaseous-looking bubble, apparently with oil in it because it was streaked with rainbow colors, welled up, this time forward of midships.

Captain Henry Nelson headed instantly for a near-by coral reef. He rammed the reef and the bow of the *Coolidge* drove well up on it and stayed

175

there. It seemed as if the cargo had been saved and, possibly, the ship. But the *Coolidge's* bottom was torn open and water was pouring in. Because of the slope, the water all rolled back into the stern. In about an hour, the stern had become so heavy that it pulled the bow off the reef and a few minutes later the ship turned over on its side, then turned topside down, and then sank stern first, the bow going down last.

The plane with this reporter in it turned around and sat down immediately. By the time I returned to the scene in one of the rescue boats, the work of life-saving was well under way.

There had been no panic, not even for an instant, survivors declared. The moment the mine had hit, the troops on board had been ordered back to their quarters to get them out of the way of the merchant sailors who were struggling to salvage the vessel and, later, speed up the work of rescue.

Cargo nets were thrown over the side to facilitate descent and the men were called up out of their quarters in sections as rescue boats were available for them. It was a nervous time for the men locked in below with grating, grinding noises going on all around them, and the sound of running feet overhead, shouts, commands, and so forth, and the ship listing and sliding occasionally. But America's

176

army men seem to be made of special stuff, for their youthful nerves held up without breaking. While the men sat waiting for their time to be called topside, and wondering if that time would come soon enough, many entertained themselves with singing — frequently racy songs. And colored troops particularly took occasion to bang out the last bits of music on cherished instruments they would have to abandon for lack of space on the rescue boats.

One of the surprising features of the incident, according to Major Costain, was the great number of army men who did not know how to swim. These men clambered obediently down the cargo nets, but when they had to let go and drop into the water, their minds just froze up and they could not release their holds. Officers climbed down the cargo nets after them and walked on their fingers to break their holds, and while this was going on, the scene took on a nightmare quality — the silent, tense men, eyes closed, faces all locked up, clinging desperately to the cargo nets, while equally desperate officers shouted at them, and walked on their fingers, and kicked at their fingers in an effort to save them, all this to an accompaniment from the interior of the ship of youthful voices lifted in ribald songs, and twanging guitars and a bedlam of phonograph records.

177

The rescue boat that carried me to the ship's side found one young soldier clambering down a rope which was fifteen feet short of the water. He was very calm. He held to the rope with his two hands and looked down at us calmly. 'Jump!' the coxswain shouted. 'I can't swim,' replied the soldier. 'Jump, we'll catch you!' we all shouted. 'Well, I don't know,' the soldier said, 'I can't swim.' A whole stream of profanity was directed at him and he swung there gently, listening us out, apparently too polite to interrupt. Then he said, 'Well, all right, but I can't swim a stroke.' Then he began to count. 'One,' he said, 'two, three,' and paused and said, 'Well, here goes,' and counted, 'four, five, six, and one for good measure.'

When he got to nine, he let go and hit just off our bow and sank like a stone. We waited, boathook ready, for him to come up. It seemed he never would come up, but finally he broke water and we hauled him on board. Then we found out he had jumped with a fully loaded cartridge-belt around his waist and with this weight on him had just plummeted on down. When he revived, spluttering, he said, 'I told you fellows I couldn't swim.'

The officers were the last to leave the ship. When the order came to withdraw to quarters, the officers retired with their men and passed their men on out to topside in orderly fashion, and shepherded them into the boats.

The one army officer who died had withdrawn to a hole with his men. By the time their turn came to leave, the only way out was hand-over-hand on a rope. This officer was the last to go, and, when he tried, he discovered he was too exhausted to pull himself up. He started to call for help.

His voice was not a strong one and there was a lot of noise as the last men aboard were scrambling off. But his lifelong friend, Captain Covill, of the United States Army, who had been searching for him and hoping he had got away, heard him and came sliding to the rescue along with a warrant officer and a defense worker who happened to be on board.

These three men stood on the edge of the hold and tried to haul up the officer who was trapped below by his own exhaustion. The last rescue boat was bobbing up and down uneasily alongside the vessel, shouting to the three men it could see, to hurry, and warning them there was not a moment to waste. But the three could not bring themselves to abandon the trapped officer.

They told the Captain to hold on to the rope and they would pull him up. They got him about halfway up, but he was too tired to hold on any

179

longer and let go and fell to the bottom of the hold.

Then the rescue boat, realizing that it would be sucked down with the suction of the ship, gave a last despairing warning. The three men did not even turn around. Captain Covill said he heard all the warnings clearly and that they all heard all the warnings, but felt abandoning this captain was more than they could do. The rescue boat pulled away hastily with a clanging of the coxswain's bell and a churning and putt-putt of the exhaust, and from where I was I could see the three small, dark figures turn for a moment to watch it go off and then turn back to the hold. Hundreds of men were watching the scene in silence from safe vantage-points, and it filled their eyes and filled their hearts to breaking.

An extra rope had been found somewhere in the hold and the Captain down below there had tied it around his waist and was trying to splice it to the rope held by his would-be rescuers and get pulled up that way when the *Coolidge* went down.

Captain Covill said that when the water hit him and boiled up all around him, he knew he was going to be sucked down with the ship, and all he could think of was to get it over as fast as possible. He opened his mouth wide and gulped in sea water.

180

Then a remarkable thing happened. The three men outside the ship were sucked down with it and went down and down, perhaps as far as to the bottom, but somewhere along the way something like an air bubble formed and threw all three to the surface where they could swim to safety. The Captain, trapped in the hold by his own weariness, died there.

CHAPTER 12

Americans Can Fight

★

CHAPTER 12

★

Americans Can Fight

Our fellows look very calm in battle. I never saw any of their faces break all up the way an actor's does, or go into sculptured lines of resolve, or anything like that. They're just there doing a job with this gun or that, or in a hole somewhere, and that's the way they look, like fellows preoccupied with a job and wrinkling up their brows over it.

This is important. The Japs are the toughest enemy we have ever had to face. As far as I could tell in those proving grounds in the Solomons, we are beating them in every department of war. This is a statement with which no ranking officer will agree publicly. They think such news given to Americans will make Americans soften up and throw away a victory there is now every reason to believe we shall win some day — not, perhaps, soon, but some day.

In the Solomons, two-to-one odds against us have been the minimum, and the odds have gone as high

185

as twelve and fifteen to one. But, just the same, we've been winning out there all along. In the five battles of the Solomons, the least we have done is keep the Japs from winning — which is victory, in a military sense when a long, hard war is still in its preliminary stages — and in our biggest successes, in the fourth and fifth battles, we not only have kept the Japs from winning, but have made them pay heavily for trying to win.

We've licked the Japs on land, on sea, and in the air. We've shown that we have more military brains than they have, are better at war, all kinds of war, from strangling and knife-fighting and head-trampling on up into the complicated mechanized operations of modern battle. The Solomons haven't shown yet that we can outproduce the Japs, but we think that's true, that we can make as good material as anybody and can make more of it than the Japs, and can replace it faster than they can.

But there's one thing that nobody in the world can be better at than the Japs and that's in the guts department. They have more guts than the Germans have. At least, they have shown thus far in the Solomons deal, which is the first deal where they've had to hold their chins out and take it, that they have more guts. The Germans have said 'Kamerad' in the past and may be relied on to say it in the future. But the Japs have never surrendered, never *en masse*, and only rarely as individuals. We have not yet taken a single officer alive on Guadalcanal, although we have tried in every way we know how. And the great majority of the few soldier prisoners we have taken have been wounded and in a condition where their minds have not been up to par.

Every day I was there, the Jap gave new evidence of his intense willingness to go to any lengths to win, or, if unable to win, to go on fighting until his breath stopped.

Under the heading of going to any lengths to win, the following incident may be cited as an illustration. The Jap seems to think it useful in land fighting to put snipers in our rear to harass us. Once, early in November, our fellows, working their way west of the Matanikou River, were held up for a day and one-half along a narrow sector. They drove the Japs out of that sector about dawn of a Wednesday and held there all that day and the next day. Toward five o'clock Thursday afternoon, a Marine, deciding to dig in for the night, found some soft-looking dirt on the edge of a tree and with the first poke of his shovel hit a Jap body. The Jap was covered over very lightly with a sprinkle of dirt, but his uniform had made him look only like some leaves and rotting

187

twigs lying amid the dirt. The Marine uncovered the Jap and through the whole brushing-off process the Jap did not move except as pushed and jostled. But nobody who knows anything takes chances with the Japs any more. So the Marine picked up this Jap's arm and let it drop. It dropped limply and the face remained motionless and emotionless as in death. The Marine did it again, half-heartedly this time, very sure that this was a dead Jap. But the Jap, who had performed the superhuman task of lying under our feet feigning death for a day and a half just in order to get behind our lines and snipe at us, proved to have a human touch around his eyes. This second time he couldn't stand it any more and one eyelid twitched nervously.

Under the heading of willingness to go on fighting, this story may be told. I haven't my notes with me, and I can't remember this Marine captain's name, but everybody called him Wimpy. Wimpy was out on patrol and ran into some Japs holed up in a native hut. Quite a hot little brush followed, and after about fifteen minutes our side got no more answering fire.

Wimpy crawled up close and saw that all the Japs were dead except one, who seemed badly wounded. This one was lying on the floor of the hut in a corner farthest from the door. He was bleeding from the mouth and stared solemnly at Wimpy, and Wimpy decided to try taking him prisoner.

For twenty minutes, Wimpy cajoled and begged and tried everything he knew, waving a handkerchief as a flag of truce, offering 'pogie bait,' as the Marines call candy, as a bribe. The Jap did not answer. The blood flowed steadily from his mouth and his face occasionally broke under pain, but he just stared solemnly at Wimpy.

So the Marine captain decided to go in after the man. He went in the door, holding his revolver in his hand, and stood there pointing the revolver. He stood as far away as he could because wounded Japs, so hurt they could not throw a grenade, have been known to pull the pin as somebody comes near them and blow up the reckless one as well as themselves.

So Wimpy stayed as far away as possible and pointed his gun. The Jap lifted himself to his hands and knees and began to crawl toward a dead Jap officer who was wearing a sword. 'Don't do that!' cried Wimpy, 'I'll have to shoot you.' Wimpy didn't dare go near the man. All he could do was point his gun and shout. The Jap kept crawling slowly for the sword and took out the sword and Wimpy stamped his foot and shouted, 'You damn fool! Oh, you damn damn fool! I'll have to kill you.' Then the Jap lifted himself to his feet and lifted the sword over his head and started for Wimpy, and Wimpy had to shoot him dead.

These are not exceptional cases. They are typical. So there can be no question of our being better fighters than the Japs. The best anybody can possibly do is be as good, and rely on our superiority in all other departments of war to give us the victory in the long run.

It's not easy to be as good. And it's important that we should be, because if we aren't we're going to lose this war, or, if not lose it, make a compromise peace which will turn over to the next generation the job of winning it. Our fellows fighting have to be as tough and the people back home have to be able to stand the losses, stand all the terrible sorrow and misery that the dead leave in their wake, and have to be able to feel that the dead husband and dead lover and dead son have not died for something that we could do without, but have swapped their lives for something worth the price. And they have to be able to keep on feeling it steadily every day for all the time it will take to win.

Our losses have been very small thus far. That is because we have been on the defensive in the Solomons since the day we took the place. The Japs have had to come after us. Soon we'll have to start north and go in after them. Then our losses are very likely to increase. There are a lot of people better able than I am to guess how the people back home are going to stand up under that. What I can say is how our fighting fellows are going to stand up under it because I've seen them do it.

In every battle I've watched out here, our side turns up with quite a few heroes, fellows who do more than they are supposed to do.

But heroes don't win wars. They help, just as everything else that's good helps. But the heroes are the exception, and it's the ordinary, run-of-the-mill guy who doesn't feel tempted to do more than his share who has to be relied on to win for our side. This doesn't sound glamorous, but I think it's accurate. So all the words that follow will be devoted to the ordinary fighting man on our side and how he measures up to this most difficult job in our history — being as tough, man for man, as the Japs.

As I have noted previously, our fellows look very calm in battle. The look of them is very provocative because, as I know from personal experience, a part of the mind seems to run away during battle and keeps trying to make the rest of the mind run away with it. The conscious mind, which is the part of the mind that knows all the things it has been taught, wants to stay and do the job it's supposed

to do, and this other part of the mind, the subconscious part, doesn't want to know from nothing
and just wants to get the hell away from there. The
subconscious mind can't go off all by itself. It has
to take you along with it. And to a fellow in this
position, particularly when the battle is long and the
struggle in himself is prolonged, it actually feels as
if the subconscious mind is laying rough hands on
the rest of him and pulling, hauling, and screaming
at him. And I know in my own case whenever I
have been in an action there has always been this
uproar in my head, this steady, wild-eyed, wild-
haired screaming, making it very difficult to think
about the work I had to do there.

But our fellows, filled inside with this demented
uproar and hemmed in all around outside by the
uproar of battle, just stay there and do the work
they have to do. They don't look like actors being
brave. They look mostly like fellows working.

Nobody looks young in a fight. I've seen lots of
twenty-year-olds out there in the middle of all that
stuff flying around and some eighteen-year-olds,
but I never saw anybody who looked much under
forty while the fight was going on. That's one way
our fellows show what they're up against. The
blood in their young faces gets watered with a kind
of liquid of fear and takes on that blued-over color

of watery milk. Their skin looks clothlike, with the texture of a rough, wrinkled cloth.

Then, when things get really thick, like when fellows start getting hit and dropping and crying out with pain all around you, and you can't pay any attention to that, but just have to keep on working your little gadget, pressing that little button or turning that little wheel or adding up that little set of figures, whatever it is — well, when it gets like that, still the faces of our fellows don't show what an actor's would.

Sometimes the flesh around their mouths starts to shake as if they were whimpering, and their eyes . . . you can see their eyes coated over with a hot shine as if they were crying. But they go right on doing what they have to do. The bombardiers keep right on figuring with their pencils on little white scratch-pads of paper, right in the middle of all that's going on, marking down figures and adding them up or subtracting or dividing and checking the answers, just like in school, looking — except maybe for the crying in their eyes and the whimpering around their mouths — all puckered up with thought, too, like earnest students in a school. And the gunners and radio operators, all the technicians and specialists, the plumbers, mechanics, cooks, the skilled laborers, doing the work of war — they're the same. They don't look tough, even when they're being their toughest, but they stick to their work and think about it, and do it, and do it well.

It's not so hard for a man to be tough, even these man-children who are fighting the war with us, if he can think, 'Well, now I have to be tough,' and get himself some internal help that way. He can put a tough expression on his face like an actor, and have something to live up to, have like a flag on his face to march along behind or a feeling of pride in himself and in the way he's behaving to keep him up to the mark. But in the actions I have seen, there hasn't been any time to think about anything like that and the fellows had to be tough without any extra help at all. They had to go on being tough, not because of something put into them by their own minds to hop them up, but just because of something that was in them already, that lives in them all the time and is there waiting to be used as needed. And all our fellows have this toughness — well, say, ninety-nine per cent.

The officers are tough, too, elderly men with paunches on them and wattles and dewlaps and so forth. They're all tough, not only in battle, where excitement carries a man along, but in the desperate, dirty, inglorious jobs done along the lines leading up to the front. The staff work, for instance, where men who have believed all their lives (as all those who are convinced about democracy believe) that men are more important than property, have on occasion to throw over their beliefs and value such property as ships and planes and battle real estate over their own men. They have, for instance, to add up the probable cost in ships of saving drowning men, and if the cost in ships is likely to be prohibitive they have to order the ships out of there, and leave the men to drown. Everybody involved in this operation finds it loathsome, and finds it leaves something in the mind that is not easy to put down. But we have done it where necessary and we shall do it again where we have to because our fellows who are doing the fighting, the elderly ones and the young ones, believe there is nothing more important in life than winning the war, and that nothing and nobody, no personal tragedies or disasters or selfishness, or any human emotion at all, can be allowed to stand in the way of winning. And in the Solomons, our fellows have shown they are tough enough to act on their beliefs.

Colonel Arthur, who was directing Marine operations west of the Matanikou during the drive that began November 1, remarked, 'The Japs are good fighters, but poor soldiers.' He was referring not only to their tactics, but to their officers' habit of wasting their men's lives on some intangible and inscrutable objective.

Nobody on our side is criticizing this habit of the Japs. The more lives they waste, the sooner the war will end. But our side is not wasting any lives. There is a line between wasting and spending lives. In the Solomons our officers have drawn this line. No life is wasted there, but our officers have had the guts to spend lives, their own as well as their men's. Which is the way wars are not only fought, but won.

It's just guessing to say what it is that lives in our fellows all the time and gives them the strength to be as tough as the Japs. But I think it's pride in being an American.

Nobody I knew ever thought much about being an American, or would fail to be embarrassed by mention of it, or, in fact, would fail to describe all this as a 'crock of hock.' But it seems to me now, after all those red-letter days in the Solomons, that it's the truth that Americans subconsciously, and without really knowing it themselves, are very arrogant about being Americans.

All the different peoples of the world involved in this world war — the primitive blacks, for instance, on Guadalcanal — have stood up very well in it. Americans seem to be aware, without knowing it consciously, that they are a mixture of all the peoples in the world, a well-nourished, athletic, free-to-think, rather less frustrated and so somewhat better integrated mixture.

Most of the fellows now doing the fighting have taken a lot of abuse in the recent past from the newspapers, and orators, and mostly the Republican politicians against the New Deal 'pampering,' and other worriers who were trying to scold and abuse and insult them into being tough about the depression, or food rationing, or soldiering, or what-the-hell.

But all those words and insults and restrictions and the whole terrible kicking around these young men have been taking from life in general ever since they were born, some of them, doesn't seem to have convinced very many of them that there is anything on earth, not any people or anything done by any people or invented or manufactured by any people, that can keep an American from doing a job he really knows he really has to do.

The overwhelming majority of jobs in this war are the dullest, most tedious kind of drudgery. They have nothing to do with fighting, but are just dangerous and crazy dull. The transport pilots, young gingery fighting cocks, flying supplies into Guadalcanal and wounded out of it, call themselves 'flying truck-drivers,' and feel quite poignantly that they've been sidetracked out of what they want to do — which is to fight or bomb.

The third time I flew into Guadalcanal, I went up on one of those transport planes. This one was piloted by a young rosy-cheeked lieutenant named George Wamsley. George started out to be a bomber pilot and then they threw him into this, much to his regret.

This was his first trip up to Guadalcanal and he was supposed to go in convoy because the weather was bad for navigation and there were plenty of Japs around. But, owing to some kind of mixup down at this New Hebrides base, the only load waiting for Wamsley at the take-off was four stretchers. He kept grumbling about sending an 'empty truck' up to a place that needs more of everything there is, and, when I pointed out that the convoy would have to leave on schedule whether he was ready or not, and that if he went up without a convoy he'd be in plenty trouble, he became intemperate and said, 'What the hell kind of a load is that for a truck — four empty stretchers and one goddamn newspaperman!'

So George stalled in one way and the other and finally a load of surgical instruments came tearing up to the tee just as the convoy was taking off.

198

By the time we were loaded up, the convoy was a hundred and fifty miles away. We never did catch it, and we had a very sweaty five hours going into Guadalcanal by ourselves, but George and his crew were all so glad to be able to do some good that they worried hardly at all, even when we wandered into the wrong place and the Japs started shooting at our defenseless selves.

That was one American doing a job he didn't like and didn't want, and wanted to get out of, and doing it the best way he knew how. There are hundreds of thousands of others like that in the Pacific. The fellows stuck on these tiny desert islands along our line of communications to Australia are like men condemned to a kind of Devil's Island imprisonment with the added refinement of the knowledge that the Japs might try to come in any day on a raid, similar to our Makin Island raid, and exterminate them to a man.

Their only company is themselves. Their only relaxation has to be found in themselves. Most of them can't even go swimming in the sea that is all around them because of the sharks, barracudas, tiger eels, and sting-rays. Yet in all these places there is an air of great cheerfulness and bustle and friendliness. Loneliness and boredom and anticipating Japs are much more terrible enemies for a

199

man than bullets. Yet these fellows come tearing down jovially to meet the planes, like yokels going to a depot to watch the trains go by in the old days before motor cars, and really put themselves out to help the wayfarers — most of whom they envy desperately — be comfortable in the midst of sand and bleached coral. Some of these fellows have been there nine months. They have the dirty end of the war, the really dirty no good end of it, but they're in there pitching all the same. It's easy for a man to be brave when he has to be, but for these fellows to keep hold of themselves when they know having a mental collapse or making a nuisance of themselves will release them from their jobs is something that requires real toughness.

I am very tired now, and will have to finish hastily. In the thousands of miles of war front that I have traveled these last months, I have seen nothing: (1) to make me think this is going to be a short or easy war; (2) to make me think we are going to lose it; (3) to make me feel anything but very proud of American guys and their girls.

November 30, 1942

THE END

'... this war is being fought to gain what in peacetime are known as islands and now are called "unsinkable aircraft carriers,"... the waters around the islands must be held and the lines of communications must be batted through and preserved over areas of thousands of square miles.'

New Guinea

Australia